W9-CYA-816

9635

QH   Bonner
349   Size and cycle
B6

DATE DUE			

Professional Library Service —— A Xerox Company

CHESAPEAKE COLLEGE
WYE MILLS, MARYLAND

Size and Cycle

AN ESSAY ON THE STRUCTURE OF BIOLOGY

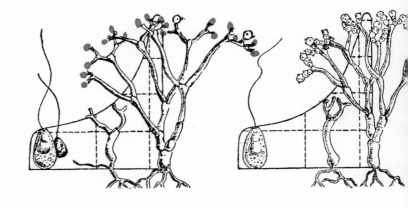

Size and Cycle

AN ESSAY ON THE STRUCTURE OF BIOLOGY

BY JOHN TYLER BONNER

WITH ILLUSTRATIONS BY PATRICIA COLLINS

PRINCETON, NEW JERSEY · PRINCETON UNIVERSITY PRESS

1965

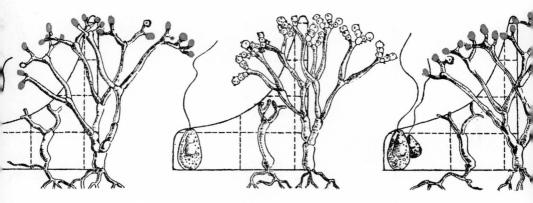

Copyright © 1965, by Princeton University Press

ALL RIGHTS RESERVED

L.C. Card 65-14306

Printed in the United States of America

By Princeton University Press, Princeton, New Jersey

Preface

THE ideas developed here began with asking questions about the mechanism of development of cellular slime molds. They turned next to questions of why this information seemed compelling, in what way was it important (matters which I have examined in some previous essays). This led to asking progressively larger questions until finally all of biology was the battleground. There it stopped (as this book will testify) and I have refrained from entering upon the even larger philosophical questions. This is partly because I am not a philosopher, partly because I think the area of biology is one in which there is some confusion, but mainly because I have a hankering for middle-sized questions. The smaller ones seem, as a complete diet, to be unsatisfactory and sometimes trivial; the larger ones of metaphysics have a way of slipping off and becoming remote from the facts, and then rushing about like self-propelled clouds. Middle-sized questions try to group and order facts, and try to make generalizations that seem both satisfying and useful to the scientist.

The progress of this book has been so slow and long that it can best be measured in geological time. As a result there are so many friends to whom I am grateful for ideas and criticism I hardly know where to end in my acknowledgments. There are certain people who showed special fortitude and patience during some of the more disorganized and incomplete early versions and I would like to express my special debt to Professor W. M. Elsasser, Mr. I. N. Feit, Dr. J. L. Kelland, Mr. W. P. Milburn, Professor J. M. Mitchison, Professor A. B. Pardee, Professor G. L. Stebbins, and two anonymous readers. I also received helpful comments from Mr. H. S. Bailey, Jr., Dr. D. R. Francis, Dr. F. E. Whitfield, and Dr. Sidney Smith. Finally, I am grateful to Professors J. O. Corliss, R. D. Milkman, P. W. Winston, L. B. Slobodkin, M. M. Swann, and Mr. A. M. Young for providing assistance on specific points.

While this book has been an occupation among others for some time, I was especially fortunate in receiving a Senior Postdoctoral Fellowship of the National Science Foundation in 1963.

The majority of the leave was spent at Cambridge University where I was able to have enough uninterrupted time to begin to find some system in masses of notes. Most of the work was done in the study of the ancient house we rented in Sawston (Where Angels Fear to Tread) which I look back upon with considerable nostalgia. I am particularly grateful to Dr. B. M. Shaffer, Dr. Sidney Smith, and other members of the Department of Zoology, as well as the Fellows of Gonville and Caius College for their kind hospitality during that period, and to Professor V. C. Wynne-Edwards of the University of Aberdeen during our short stay there.

I should like also to acknowledge my gratitude to Miss Patricia Collins who, with painstaking care, and sensitive artistry, has brought back the ancient craft of the copper plate into scientific illustration. It has been a pleasure working with her, and we are both grateful to the advice and encouragement of Mr. P. J. Conkwright and Mr. Marshall Henrichs who were kind enough to oversee the artistic aspects of the project.

Finally I acknowledge with pleasure a special grant for the preparation of the illustrations from the National Science Foundation through a grant administrated by the Research Board of Princeton University.

JOHN TYLER BONNER

Margaree Harbour
Cape Breton, N.S.
August 1964

Contents

9635

Size and Cycle

AN ESSAY ON THE STRUCTURE OF BIOLOGY

1. Introduction

FOR some years I have been concerned about the problem of the relation of genetics, evolution, and development. It has always seemed that their conventional relation is to some extent static and contrived, while in fact they must be closely integrated and part of one scheme. A problem with so many interrelated facets can be looked at a number of different ways, and here one will be chosen which it is hoped will help to reveal some of the fundamental relations. The method holds no dispute with any of the basic tenets of modern biology; this is not an attack on either the facts or the theories of biology today. Rather it is a regrouping of those facts and theories in such a fashion that new and deeper insights may possibly be achieved.

Part of our present difficulty is that so many of our ways of looking at biology have slowly grown out of the past that we are infiltrated and imprisoned by a massive tangle of traditions and conventions. What we teach today is part biology and part history. There is nothing wrong with this but we do not always know where one begins and the other ends.

The view taken here is that the life cycle is the central unit in biology. The notion of the organism is used in this sense, rather than that of an individual at a moment in time, such as the adult at maturity. Evolution then becomes the alteration of life cycles through time; genetics the inheritance mechanisms between cycles, and development all the changes in structure that take place during one life cycle.

When looked at this way the size of the organism in the cycle takes on a particular significance. Size is correlated with time, for in general large organisms have long cycles and furthermore the different parts of the cycle can be readily classified on the basis of their size characteristics.

The life cycle is a summation of all the molecular or biochemical steps, one following another in a well-ordered sequence. The difference between two cycles is a difference in the nature of the steps. They differ in their structure, their composition, and therefore the life cycle is a qualitative unit. A change in size of the organism, on the other hand, does not necessarily require that there be a qualitative difference in the steps, but merely more steps. Therefore the life cycle is quality, and size is quantity. One is a statement of composition of matter and the changes in that composition; the other is merely a statement of quantity. Just as in chemistry it is vital to know for any reaction the nature of the ingredients as well as their amount, in biology we are concerned with size and cycle.

In this book biology will be examined within this frame, and the question may fairly be asked what are we to gain from this approach? What will we have that we do not have at the moment?

I hope first that we will see more clearly what are the prime questions to be answered by the experimental biologist so that the mechanism of the forward progression of the life cycle may ultimately be understood. Secondly, I hope that we will see more clearly what are the prime questions to be answered by the environmental or evolutionary biologist so that the mechanism of evolutionary progression and ecological success may ultimately be understood. Finally, and most important of all, I hope that it will be possible to show that these two great areas of future inquiry and research are not separate, but firmly woven together into one large fabric; biology is not two disciplines, but one.

It may be helpful to examine in more detail how the view taken here contrasts with the traditional one. Even though the same thing is merely being said a different way, the difference is important and should be understood clearly.

In standard biology the classification of organisms is based primarily on the following four criteria:

1. In the first place animals and plants are put into groups on the basis of their structure or morphology. The principal criterion for groupings such as species, genera, and all the larger, more inclusive groups is structure. This is particularly obvious in the case of asexual organisms, but in all cases it is appreciated that differences in structure have a genetic basis and that differences between species must be thought of as differences in gene pools. But the objects that are identified and classified are the construction and the shape of the parts of an organism. This is the basic

currency of modern biology, though it is understood that this currency has genetic and evolutionary significances that are not always stressed, just as when one mentions a sum of money one need not mention where it came from or how it will be used. The objects of biology, organisms, are identified and classified on the basis of morphology.

2. An organism is traditionally identified as an adult. Many authors have pointed out this curious fact. Although some organisms are classified at least partially on their larval characteristics, by and large the object that is classified on the basis of its structure is the adult. As de Beer (1958) and others have pointed out, this has been a matter of necessity in paleontological studies, because usually it is the adult alone that is preserved, but the tradition is far more deep rooted than this. From the very earliest times, from Aristotle, from Linnaeus, and many others we find the structure of the organism automatically equated with the structure of the adult. I would suspect, for instance, that the average reader would assume that in the preceding paragraph I was referring to adults only, even though I do not explicitly say so. This is so firmly engrained in our way of thinking that it can only be consciously repressed. Perhaps it has something to do with the fact that we ourselves are adults, and it is difficult not to be subconsciously self-centered. Perhaps it is merely a matter of convenience; it is in fact the most practical solution to the problem. Whatever is the reason, it is a deeply set tradition.

3. Since Darwin and Mendel it has been recognized that variation is essential for evolution and natural selection, and furthermore it is fully appreciated that sexuality is a prime means of providing variation. As a result, sexuality has been raised on a remarkably high pedestal and is generally assumed to be the only source of variation (even though many biologists know and have argued otherwise); sexuality has become a prime basis for the classification of organisms. This has occurred in two ways. One is rather artificial in that the sexual parts of an organism, such as the construction of the flower, are often used as systematic criteria. The other is that in sexual organisms the inability to interbreed is considered the ideal criterion for the identification of two populations as separate species. Again, if we turn to explanations, it is hard to see why sexuality has become so conspicuous

in our system of classification, although this is a problem for which Freud might provide a ready answer.

4. Finally, besides being structural, adult-centered, and sexually oriented, our traditional system of classification is basically concerned with phylogeny. Again this is not a new point, and to counter it many biologists, especially ecologists, have been interested in convergence and in functional adaptations to a particular environment. But on the whole blood relations have been far more important than structural-functional similarities. For instance the fact that the wing of a bat is related to the forelimb of quadruped through descent is considered of greater significance, than the fact that the butterfly, the bird, and the bat can all fly with quite different structures which have had different evolutionary histories. Homologies are the basic stuff of biological classification; analogies are intriguing curiosities. In this notion, more than any other, we see our strong desire to bind history and biology. Evolution can only be thought of as a process in abstraction; our first desire is to see the phylogentic tree all neatly laid out so that the begats go back to the first moment of life on earth.

This is partly because natural selection, and our whole conception of evolution remains the most encompassing and the most useful theoretical frame that exists in biology. All the four aspects of classification that we have discussed are held in place by the firm grip of natural selection. Structure is the object of selection; the adults are adapted, by selection, to a particular environment; sexuality is the means of variation so necessary for selection as well as an isolating barrier in the formation of species; phylogeny is the life tree that is shaped by natural selection.

In modifying these four points, as I shall now do for the proposed alternative method of organizing biology, it must be emphasized that I have no quarrel with the significance of natural selection, nor with its grip on any scheme of classification or organization. My point will be simply that the traditional view is too restricted, and that a broader one will more accurately reflect the true state of affairs. But by taking this broader view something is lost as well as gained. The old system is illogical partly because it is useful, and what the new system gains in inner consistency it loses on the practical side. It is far easier to classify adults rather than life cycles. Adults can be skinned and dried, stuffed in drawers with neat labels attached; but storing a life cycle is a relatively impossible task. Even the whole approach to the problem of the identification of species is practical rather than logical, for as Mayr (1963) points out, in sexual forms one uses the criterion of the lack of interbreeding, while in asexual forms one uses morphological criteria. Nevertheless, though it may be impractical and cumbersome in the field and in the laboratory, a broader perspective may in fact be simpler and may shed more light at the source of all the problems. Let us now reconsider each of the four points mentioned above.

1. Along with morphology or structure, let us add size. As has already been mentioned, and will be stressed in more detail later, structure is quality, while size is quantity. But in a flash the sensible, practical biologist will say what nonsense, for how can one classify animals on the basis of size? In any one genus, for instance, one may find an extremely large range of sizes and therefore this could hardly be a useful criterion to distinguish genera. One can only agree, and I would not ask the systematist to use size any more than he now does. But this is not the point here, for I do not wish to use size to characterize taxonomic groups but rather to characterize particular life cycles. Size may be an important difference between two species in one genus and have consequences which permeate into its ecology, its reproductive activities, its evolutionary progress, its development, its physiological activities. In fact size is as important as morphology, or to return to our original words, quantity can have as much significance as quality.

2. Not only the adult, but the whole life cycle will be considered the organism. This is an ancient notion, for philosophers have often pointed out that an *individual* conventionally means an organism in a short instant of time—in a brief time-slice. For example, if we refer to a "dog" we usually picture in our minds an adult dog momentarily immobilized in time as though by a photographic snapshot. The philosopher is quick to point out the shallowness of this convention, for is the dog not a dog from the moment of the fertilization of its egg, through em-

bryonic and foetal development, through birth and puppyhood, through adolescence and sexual maturity, and finally through senescence? As a matter of fact even the decaying carcass of a dead dog is still a dog, although we imagine ourselves clear in our minds as to the difference between the alive and the dead. Therefore we can say that in a more general sense an individual live dog may be defined as that which fits in the whole span between fertilization and death, although later we shall consider the problem of a precise definition of a life cycle.

Stress on the whole life cycle has already been provided by the developmental geneticist, for he has seen the great significance of the fact that genes do not express themselves just in the final stages of the formation of the adult but all during the course of development. Furthermore it is well recognized that certain genes do not act until senescence, and therefore the thought that genetic effects appear during any stage of the life cycle is quite accepted. The ecologist also has been aware of the significance of the different stages of the life cycle. But here we are making the even more radical suggestion that for all considerations the life cycle as a whole is pivotal.

3. Variation control can be achieved in a number of other ways besides sexuality, a matter which will be considered in detail later. Though this is true, it must be admitted that sexuality is by far the most important method. Since we are not concerned here with traditional taxonomy and species identification, its significance there is outside the present discussion.

Sexuality frequently plays a key role in the discussion of life cycles, but again it will not be emphasized here. For instance the position of meiosis in the life cycle will determine what part of the cycle is haplophase and what part diplophase, but whether the genome is represented once or twice (or more) seems to have remarkably little effect upon the construction of the organism itself. For instance, by experiment it is possible in the moss to prevent fertilization and produce haploid sporophytes without any significant morphological change. And again the ploidy can be changed by experiment in many plants and animals without radical structural modifications. Because of these facts, and because asexual organisms also have life cycles, the number of chromosomes in any part of a cycle is unimportant as far as the progression of the cycle itself is concerned. As will be shown, its significance lies rather in the control of the amount of variation; it is one of many factors which contribute to this control.

4. It has already been stressed that here we are not interested in phylogeny or practical systematics, and that the system of classification to be used is based on the life cycle of organisms. Therefore analogy rather than homology, and convergence rather than ancestral ties, will be emphasized. The characteristics of a particular cycle are significant, rather than how they arrived through history. This is not to say that the historical information is neither interesting nor important; it is simply not of concern here. We do not care whether a particular functional adaptation is convergent or phylogenetically related; rather the nature of the functional adaptation will be examined. It is therefore not so much a question of how it got there, but what it is.

The only thing that will not change in our comparison of the two approaches to the organization of biology is the role and significance of natural selection. As before, everything lies within its firm grip, but now instead of saying that merely structure is the object of selection, we add that quantity or size is also. Before we said that adults were adapted, and now we add that all the other stages of the life cycle are adapted too. Before we said that sexuality was the means of producing variation for selection, and now we say there are other lesser means as well. Before we said that phylogeny was at the center of all systematic considerations, but in our new systematics the immediate existence of a life cycle, with its functional adaptation to its surroundings and its corresponding structural changes, is of prime importance.

So far, all I have done is to point out some of the main features of the approach to be used here and contrast them with traditional ones. Now we must delve more deeply into the approach and examine the method in detail.

2. The Method

Definition of the steps

One of our main concerns in discussing the method is to have clear definitions of both life cycle and size in the senses they are to be used here, but even before this can be done it is necessary to describe and define a more basic concept that is common to both.

All biological processes are made up of a series of chemical reactions which follow one another in a sequence. For instance, in a physiological process such as muscle contraction, or in the transmission of a nerve impulse, there is a chain of chemical reactions that fire off in a series. These reactions are invariably dependent upon the structural positioning of the reactants as well as their existence, and thus the process is carried out in a controlled fashion.

In developmental processes, or what might better be called life cycle processes, the number of chemical reactions that takes place is much greater; in fact the sequence may be enormously long when one thinks of all the reactions that must take place within a multicellular organism between fertilization and death.

If one now adds all the life cycle generations that must occur for evolutionary change, then the number of chemical reactions that occur becomes great beyond imagination.

For convenience we shall call these chemical reactions *steps*. They are characterized by being in set and predictable sequences, one set of conditions leading to another and so forth. For the purpose of definition, *the steps are chemical reactions that occur in a definite sequence in time within living organisms*. These steps tend to recur in repeating patterns: if the cycle of repetition is short and involves few steps it is a physiological process, such as is involved in nerve transmission and the subsequent recovery and repolarization of the fiber. If the steps are numerous before repetition then we are on the level of a life cycle. There are certain physiological processes, such as hormonal changes, which are intermediate in their duration and number of chemical events.

There are two important unrealities about this way of abstracting biological processes. The first is that there is the implication that all the chemical processes are roughly similar, both in their complexity and in their duration. This of course is absurd; there is a tremendous range in the kinds of reactions that occur. Momentary reflection on the processes that have been exposed by biochemists makes this evident. The second point is that most biological processes, and certainly life-cycle processes, do not consist of a single chain of steps or chemical reactions, but of many going on simultaneously. In fact the larger the organism, the greater the opportunity for simultaneous series of steps. These series may periodically interlock in set ways, or they may not, but they undoubtedly will be multiple.

With these qualifications clearly understood, it may be useful to represent pictorially the sequence of steps as a single series of arrows of equal length. For instance the steps from fertilization to maturity might be shown thus:

fertilized egg $\rightarrow$ $\rightarrow$ $\rightarrow$ $\rightarrow$ $\rightarrow$ $\rightarrow$ $\rightarrow$ adult
(or equivalent)

If the adult now reproduces and produces a series of offspring that join the adult in a family or a population we could make a further addition.

fertilized egg $\rightarrow$ $\rightarrow$ $\rightarrow$ adult $\rightarrow$ $\rightarrow$ $\rightarrow$ population.

Again it must be emphasized that the single series of arrows does not mean that the population is produced by one sequence of steps, but of course many parallel ones. In fact the population is made up of many interbreeding adults which, were it to be represented accurately, would give a maze of interweaving lines of arrows. These simple diagrams may be useful in making some points presently, but their symbolic simplification must always be kept in mind.

How natural selection can change the steps

In order for natural selection to occur, there must be a modification of the steps, for the sum of the steps are the organism, that is, the life cycle. This modification is variation and it is usually conceded that the majority of variation is controlled by the nuclear genes, where mutation and recombination occur in gametogenesis and fertilization. This is the point at which *innovations* of this sort may be introduced. But natural selection also involves the *elimination* of certain variants so there can be progressive change in the constitution of the population, and this of course occurs by differential reproduction; certain adults are more successful in flooding the population with their offspring than others. Therefore we can consider the point in the life cycle when reproductive maturity is achieved as the point where it is determined which variants succeed and which are eliminated. We are not saying that only adults die, but rather that selection eliminates particular genes by blocking the number of progeny which harbor them. The blocking itself can obviously be achieved many ways and at different moments during the life cycle; but the end result is elimination of the genes by differential reproductive success. We can now attach to our previous diagram the points of innovation and elimination in the sequence of steps.

fertilized egg $\rightarrow$ $\rightarrow$ $\rightarrow$ $\rightarrow$ adult $\rightarrow$ $\rightarrow$ $\rightarrow$ $\rightarrow$ population

⇧ innovation ⇩ elimination

The stretch between the point of innovation and the point of elimination is the life cycle.

It is important to note again that physiological cycles of various sorts (long ones such as hormonal cycles and short ones such as nerve or muscle reactions) would occur on any small segment of this life cycle span.

Another fact of considerable interest, illustrated in the diagram, is that while the two pivotal points for natural selection frame the life cycle, populations lie beyond these points. This means that any genetic change which affects groups of organisms, such as genetically determined systems of communication, must be initiated and culled in the life cycle, and their effects can only affect the population by extrapolation. Much more will be said of this matter later, but here the important point is that the population changes which occur during the course of evolution are achieved entirely through natural selection acting at the level of the life cycle.

The same applies to any gene-controlled physiological process or biochemical process that occurs within the cycle; or in our terms, any one of the steps or small groups of steps. That is, a step can be changed by natural selection only through the agency of the life cycle, for only the whole cycle has the point of innovation and the point of elimination, both of which are necessary for the change.

The life cycle, in fact, is the unit of evolution, the unit of innovation and elimination, and it is for this reason more than any other that the life cycle has a central position in the structure of biology.

Size as an analytical tool

We must momentarily digress to consider the question of what we mean by size. It has already been said that size is like quantity in physical laws, but confusion may arise from the fact that molecules, cells, organisms, populations, and even communities can be large and small. The answer is straightforward, for each of these may be separately designated as large or small; there is no need to confine the word size to any special category.

There is one particular size that is of special interest here, and that is size as it applies to the life cycle. This means specifically that the organism at any one moment during its life cycle, has a particular size.

If we are to talk about size in any specific way it is necessary to be able to measure it, and this brings up a practical problem. What index is to be used to indicate the size of an organism: weight, volume, or length. After careful consideration of this question, which included such trivial matters as the availability of measurements in the literature, comparison of different kinds of organisms, and others, it was decided that none of the measures was ideal for all organisms. Therefore the simplest (and most generally available) measure, length, was chosen and will be used throughout this study, despite its limitations of accuracy. In fact it could be argued that the use of length as an index of size is less questionable than some of the other approximations used in the various correlations that will be presented.

If we first consider the question of size changes within the cycle, then it is clear that as a rule the point of fertilization (the point of the introduction of innovation) is the point where the organism is small, while the moment when the organism is capable of reproduction is usually the largest point in the cycle, or at least approaches it. If we were to use size as an index, then, we could say that the life cycle is framed by the point of minimum size and the point of maximum size, and modify our diagram accordingly:

point of   point of
minimum → → → → → → maximum → → → → population
  size   size
   ⇧   ⇩
innovation   elimination

The diagram will take on much more meaning if the reasons for the changes in size are understood. Size increase is equivalent to growth. Growth may involve making more of exactly the same thing, or it may be differential and a number of different things are synthesized. In any event the total bulk of living stuff increases by biochemical synthesis. The point of maximum size is therefore quite simply the result of growth.

The point of minimum size is achieved by the fragmentation or separation from the large reproductively active adult. This idea has been clearly expressed by Picken (1960), who considers the life cycles of multicellular organisms in terms of alternating adhesive and non-adhesive states. This is particularly striking when one considers the gametes separating from the parents as cells. Besides applying the principle to this case of Picken, we can add other types of separations that extend beyond it. For instance in certain asexual cycles such as the budding of hydra (Plate 19) or the fission of planaria (Plate 20) the loss of adhesion is extremely localized. Also we must make the proposition sufficiently general to include unicellular organisms, where there is a permanent lack of adhesiveness and the cells separate after each division (Plates 1, 2).

In all these cases there is a separation, and immediately following this separation there is point of minimum size. Therefore, just as maximum size was produced by growth, minimum size is produced by separation. It will be a basic assumption throughout this book that growth or size increase is adaptive and provides some special advantageous features at a particular period during the life cycle. Likewise, separation or abrupt size decrease is also adaptive and is concerned with reproduction.

Size as a function of time

We can also consider the size of the whole life cycle, that is, the total number of steps within a cycle. It has already been pointed out that the duration of various steps will be different, but with a given mean duration, the more steps, the more time elapsed. Thus one would expect that larger organisms would have longer life cycles, for it also takes more steps (and therefore more time) to make a large individual. Common sense tells one that it takes longer to build a sky-scraper than a chicken coop, but fortunately this can be tested empirically for organisms.

It is possible to determine the length of an animal or plant at the time of reproduction and compare this with its generation time, or time to reach reproductive maturity (Table 1). As may be seen from Figure 1, if these two quantities are plotted against each other, on a logarithmic scale, one sees a clear-cut relation; the larger the organism the greater the generation time. This relation holds from bacteria to the giant sequoia, and despite the scatter the trend is clear. It is also interesting to note that plants and animals lie together roughly along the same line.

Size and time are therefore related; the longer the period of size increase, the greater the size at the point of maximum size.

I should here acknowledge an aspect of my methodology that is bound to cause dismay to some. A number of generalizations are made to which any biologist will immediately think of numerous exceptions. But I sincerely hope that enthusiasm for exceptional cases will not blind the reader to the truth of the generalities. It is, after all, quite accepted that in a quantitative experiment, a statistical significance is sufficient to show a correlation. The fact that there are a few points that are off the curve, even though the majority are on it, does not impel one to disregard the whole experiment. Yet when we make generalizations about trends among animals and plants, such as changes in size, it is almost

Table 1. A comparison of the length and generation time of different organisms. See Figure 1.

Species	Length		Generation time		Reference*
Staphylococcus aureus	1	μ	27	min	μ
Bacillus cereus	2.3	μ	19	min	μ
Pseudomonas fluorescens	3	μ	30	min	μ
Escherichia coli	3.5	μ	12.5	min	μ
Spirochaeta sp.	20	μ	8.8	hrs	μ
Euglena gracilis	40	μ	7	hrs	μ
Tetrahymena geleii	50	μ	2.5	hrs	μ
Didinium nasutum	100	μ	6.6	hrs	μ
Paramecium caudatum	200	μ	10	hrs	μ
Stentor coeruleus	1	mm	34	hrs	G
Daphnia longispina	1.4	mm	80	hrs	BH
Drosophila melanogaster	3	mm	14	days	
Venus mercenaria (clam)	6	mm	1.5	yr	BH
Musca domestica (house fly)	7	mm	20	days	BH
Tabanus atratus (horse fly)	14	mm	110	days	BH
Apis mellifera (honey bee)	15	mm	20	days	BH
Crassostrea virginica (oyster)	30	mm	1	yr	BH
Anolis carolinensis (chameleon)	47	mm	1.5	yr	BH
Lymnaea stagnalis (pond snail)	55	mm	9	mo	BH
Rana pipiens (leopard frog)	60	mm	1.5	yr	BH
Aequipecten irradians (bay scallop)	78	mm	1	yr	BH
Diemictylus viridescens (eastern newt)	80	mm	2	yr	BH
Peromyscus gossypinus (cotton deermouse)	90	mm	70	days	M
Asterias forbesi (starfish)	100	mm	1.5	yr	BH
Terrapene carolina (common box turtle)	100	mm	4	yrs	BH+G
Dicrostonyx groenlandicus (collared lemming)	140	mm	48	days	BH
Callinectes sapidus (blue crab)	150	mm	13	mo	BH
Limulus polyhemus (horseshoe crab)	200	mm	10	yrs	BH
Ambystoma tigrinum (tiger salamander)	205	mm	1	yr	BH
Neotoma floridana (wood rat)	250	mm	7	mo	BH
Vulpes fulva (red fox)	590	mm	1	yr	BH
Castor canadensis (beaver)	680	mm	2.8	yrs	BH
Masticophis taeniatus (western striped racer)	870	mm	3	yrs	BH
Odocoileus virginianus (white tailed deer)	1.1	M	2	yrs	G+M
Homo sapiens (man)	1.7	M	20	yrs	G
Ursus horribilis (grizzly bear)	1.8	M	4	yrs	BH+M
Cervus canadensis (elk)	2.3	M	2.7	yrs	BH+M
Diceros bicornis (black rhino)	3.0	M	5	yrs	G
Loxodont africana (African elephant)	3.5	M	12.3	yrs	BH
Cornus florida (dogwood)	9.0	M	5	yrs	G
Abies balsamea (balsam fir)	16	M	15	yrs	G
Nereocystis Luetkeana (kelp)	20	M	2	yrs	F
Betula alleghaniensis (yellow birch)	22	M	40	yrs	G
Balaenoptera musculus (blue whale)	22	M	5.5	yrs	BH+M+S
Abies concolor (white fir)	45	M	35	yrs	G
Sequoia gigantea (giant sequoia)	80	M	60	yrs	G

* B.H., *Handbook of Biological Data,* Ed. by W. S. Spector; G, *Biological Handbook: Growth,* Ed. by P. L. Altman and D. S. Dittmer; M, *The Mammal Guide* by R. S. Palmer; S, *Whales* by E. J. Slijper; F, Frye, *Bot. Gaz.* (1906).
(Note: in the case of quadrupeds the length given is from the base of the tail to the tip of the snout.)

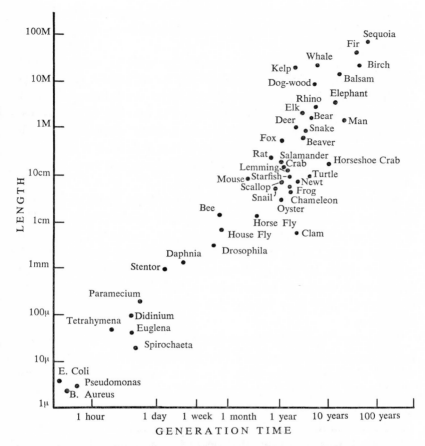

FIGURE 1. The length of an organism at the time of reproduction in relation to the generation time, plotted on a logarithmic scale. This graph shows the data in Table 1.

automatic to point out the exceptions and throw out the baby with the bath. This is not a question of fuzzy logic or sloppy thought; it is merely a question whether the rule or the deviations from the rule are of significance in the particular discussion. For this discussion the rule will be of greater importance than the exceptions.

Complexity as a function of size

Another instance where the same problem occurs is in the matter of complexity. It is often said that microorganisms are just as complex as whales, and indeed this may be true if one picks one's definition of complexity appropriately. The problem is one of definition.

In the present discussion complexity will be equated with differentiation or division of labor. To be as precise as possible we shall define it as a function of cell types (cell differentiation) within an organism, that is the larger the number of cell types, the greater the complexity. But with a moment's reflection it will soon be apparent that the precision of this definition is somewhat illusory, because the identification of different cell types is often a difficult task.

Despite these problems, an attempt has been made to plot the length of various organisms against their approximate number of cell types (Fig. 2). In this graph the organisms have been selected as the largest organisms for a particular number of cell types, so this is a curve of the maximum size possible for a given degree of division of labor; and in fact one finds many small organisms which may be, by this definition, as complex as the large ones.

The curve, however, shows that size increase is accompanied by an increase in complexity (even though one may have complexity without the large size). It is another way of demonstrating a principle which has been discussed in detail elsewhere (e.g. D'Arcy Thompson (1942), F. O. Bower (1930), Bonner (1952)) that increase in size is accompanied by many structural changes that are dictated by surface-volume relations. This has been previously referred to as the *principle of magnitude and division of labor* (Bonner, 1952).

It is also of interest to note two other facts of the approximate relation shown in Figure 2. One is that plants show a greater length for any particular number of cell types. This is perhaps reasonable when one considers their lack of locomotion and their accumulation of rigid supporting materials. The other is that as organisms become large, the number of cell types (the complexity) increases at a much greater rate than the length. Apparently size is extremely expensive in terms of complexity, for a small increase in size is accompanied by a large increase in complexity.

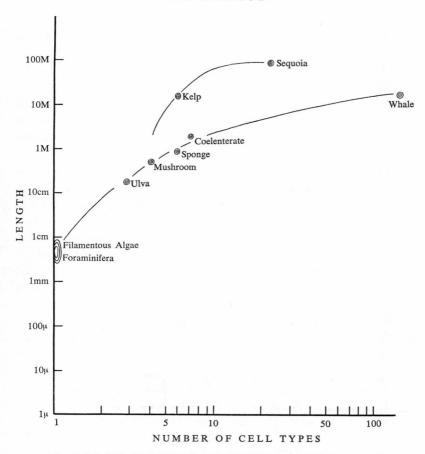

FIGURE 2. Size *vs.* complexity. Graph (on a logarithmic scale) showing a very rough estimate of the maximum size known for a given number of cell types (which is used as a measure of complexity). In examining this curve one must keep in mind the difficulty in making precise estimates of the number of cell types and the difficulty of comparing size in organisms of radically different shapes. Note that in general animals, for a given maximum size, tend to have more cell types than plants, which is perhaps related to their mobility and flexible cell construction. Also note that as one proceeds up the scale, relatively small increases in size are expensive in terms of the number of new cell types needed.

Conclusion

A number of authors, especially Waddington (1940, 1957) and Haldane (1956), have made the point that biology is in need of connecting laws that bring together molecular events, developmental (or life history) events, and evolutionary events. This, they point out, is precisely the advantage that physics has over biology, for there is a universal set of laws that applies from the internal structure of atoms to the galaxies of the universe. It would be grossly misleading to say that the method proposed here in any way provides such laws, but it does bring the three levels (molecules, life histories, and evolution) into view at one time, and at least we can see how they hook together.

By use of the concept of steps it is possible to break down all biological processes to an abstract designation which refers to the sequential biochemical reactions. The sequence of these steps has a recurring pattern which involves separation and growth or size increase, providing points of maximum and minimum size. The point of minimum size is the point at which innovations (variations) may be introduced into the genetic system. The point of maximum size is where certain variants can be selectively eliminated by differential reproduction. Between the points of minimum and maximum size we have the life cycle, and because of the properties of innovation and elimination, the life cycle is the unit of evolutionary change. This means that any alteration of any biochemical step and the fixation of that alteration in the population is achieved through the agency of the life cycle. For convenience the parts of the life cycle may be identified on the basis of size, and furthermore, the size achieved at the point of maximum size is related to the duration of the life cycle. Finally, there is a limit to the size increase that can occur without an increase in complexity in the sense of increased division of labor.

The life cycle is a connecting principle, and in some ways the most far reaching property of cycles is size. Armed with this method we may now venture forth and apply it to the facts of biology.

3. Size in the Cycle

Periods of size increase and size decrease

Our system of classifying the stages of the life cycle has already been described in its broad outline, and now more precise detail is needed.

The fact that the point of separation is the point of smallest size presents no difficulties, but there are some problems with the point of maximum size. First we must specify that our interest is in the maximum size during an extended period of the life span and not just after a big meal. The concept of maximum size must be above fluctuations in dietary successes or difficulties. Secondly the period of maximum size may not necessarily be the adult; for instance caterpillars weigh more than the butterflies into which they transform (Plate 22). But despite this and other examples of large larval or embryonic stages, in general it is the adult that is the largest.

We have already seen that reaching the point of maximum size is a slow process, while in contrast the point of minimum size is reached in one quick drastic step. A unicellular organism divides in two (Plates 1, 2), hydra pinches off a bud (Plate 19), egg and sperm are cut off from the gonadal tissue. It is the difference between building a house and knocking it down.

Other than this sudden or abrupt one there are other periods of size decrease, which are significant parts of some life cycles. For instance during metamorphosis of many animals, insects (Plate 22), echinoderms (Plate 23), ascidians (Plate 24), or amphibians (Plate 25), there may be a decrease in size which precedes a final period of size increase.

Another slow type of size decrease found in some animals is the period of senescence or decay which is often accompanied by a definite decline in cell number and over-all weight. However this period of decline should really be considered parenthetically because it is only found in some organisms, and in many of those it is only found during a post-reproductive period. In those cases where it is only found in sterile old age it is, in a sense, not part of the life cycle, but an appendage to it. However this distinction is not very sharp, for decline in size through aging, at least in human beings, is a slow process spanning both part of the reproductive period as well as the post-reproductive period, and therefore it cannot be said to be totally outside of the cycle.

This raises a further problem. If we say that development is a period of size increase and senescence a period of decrease, perhaps we could add that between these there is a period of equilibrium which we call maturity, at least in those organisms that clearly demonstrate such periods. Theoretically these distinctions are sharp, but in practice they are considerably vague. Some growth processes continue throughout the periods of maturity and even into old age, such as the synthesis of red blood cells in mammals, while some decline in the body before maturity has even been reached. Or to give the extreme case, as Minot did many years ago, senescence starts at fertilization, for the whole life cycle consists of a progressive decline in the growth rate. In any event the phases of development or increase in size, maturity or equilibrium, and senescence or decline are not sharply delineated and fuse into one another. This is illustrated especially clearly by the fact that reproduction may occur annually over many years and during any one or all of the major phases of size change of the life cycle.

Minot's law of the decline of the specific growth rate brings with it the important idea that all the phases of the life span are under a common bond; they are subordinate to some control mechanism that carries all processes through from one stage to the next. This fits precisely with the arguments that have been stressed here, for each of the phases is part of the all-important life cycle. The pattern of the whole cycle is controlled from beginning to end, and the fact that some periods involve increases in size while others involve equilibrium or decrease is all part of the plan of rigid steps of the complete cycle.

Periods of size equilibrium

Despite the imperfections and problems connected with subdividing the life cycle into parts, it is necessary to do so for further analysis. Thus far we have stressed the periods of size increase and decrease, but we have only given one example of size equilibrium or constancy. This was the example of maturity where, in a very loose sense, the constructive and destructive processes are in balance.

However there are very much better cases of stages of size constancy or size equilibrium. The most obvious are stages of dormancy. In plants this consists of a spore or a seed encased in a hard shell that can withstand all sorts of environmental hardships. As far as the life cycle of the plant is concerned, this is a rough kind of suspended animation.

There is an equivalent in animals in the form of resistant eggs or hard egg cases. The fertilized eggs of many species of invertebrates can lie dormant during unfavorable seasons and begin to develop once the weather takes a turn for the better. In some cases (as in plants) there are asexual spore-like bodies, such as the hardened gemmules of sponges.

Seeds differ from spores, since in seeds the embryo has already developed to a considerable degree before it shuts itself in for the dormant period. Some insects have a roughly parallel phenomenon in the form of pupation, when the larva builds itself a cocoon where it may lie dormant for long periods and eventually emerge as an adult. In a similar fashion some parasitic worms, such as *Trichina*, will form cysts.

Another less obvious kind of dormancy is found in many animals, and is particularly elaborate in certain mammals. This is the process of hibernation. Bats, for instance, will enter a torpid state, their body temperature drops, and their metabolic engines just barely turn over. Again hibernation (as its name implies) is a method of existing during unfavorable seasons. Botanists ordinarily do not consider a tree, for instance, to hibernate, but in a sense it does. An oak tree sheds its leaves in the fall, therefore eliminating photosynthesis, and during the cold winter months its cell respiration must be greatly reduced. This seasonal change is of course reflected in the thickness of the annual rings, which are an indication of relative activity or dormancy.

One might also wonder whether the diurnal period of sleep found in higher animals would not also be considered a kind of stationary phase. It is admittedly a weak example, but there is a daily reduction in metabolic activity, and mammals, for instance, spend a surprising percentage of their total existence asleep. But as we pass from hibernation to sleep the time interval is becoming increasingly shorter, and these cyclic processes could more appropriately be considered physiological cycles rather than key stages in the life cycle. Hibernation, unlike pupation in insects, is not an essential step in the forward progression of the cycle.

Conclusion

If size is used to identify parts of the life cycle, we find that there are three main periods; an increase in size, a decrease in size, and a period of no change or equilibrium. Furthermore we have seen that size increase is invariably slow and deliberate, while size decrease may be abrupt, as in reproduction, or gradual as in senescence, or intermediate, as in some metamorphic changes in animals. In a simple and straightforward way this gives us reasonable subdivisions of the parts of the cycle, and furthermore it is sufficiently general to apply to the cycles of all the animals and plants.

The remainder of this chapter will consist of a brief survey of different life cycles identified from the point of view of these size changes. This is done both by means of a text and a set of plates (Plates 1 to 30). It is not in any sense a comprehensive systematic treatment of life cycles, but rather a series of examples which illustrate the different patterns of size increase, size decrease, and size equilibrium among living organisms.

THE PERIOD OF SIZE INCREASE

The period of size increase in organisms may be further classified by the way in which the size is achieved. The basis of the classification is the building block unit, that is, the cell. Most cells are roughly of the same order of magnitude for all organisms, and therefore a major subdivision of size will be between unicellular and multicellular organisms. There is also a middle ground which includes aggregative organisms; in an approximate way they are intermediate in size. Multicellular organisms can be compounded into colonies as, for instance, among corals or ascidians. In this case the individuals are physically attached; if

PLATES 1 to 30

In this series of plates a variety of animals and plants are all drawn on the same scale. This has been made possible by using the logarithm of size and the logarithm of time. The upper contour of each cycle is the length of the organism (ordinate) at any one moment in its life history (abscissa). It is therefore possible to keep size and time in view simultaneously, and because the scales are logarithmic, small bacteria and giant trees or whales can be placed on the same grid and readily compared.

It should be pointed out that there are some imperfections in the method. In the first place beginning the graphs at 1μ and 1 minute is entirely arbitrary. Obviously, since the scale is logarithmic, one could continue many units down the scale and enter the realm of molecules and even smaller. But here the graphs isolate the range of sizes of cells and multicellular organisms (and in Plate 30 a social insect colony).

Another restriction is that all the drawings are done in their proper linear proportion, and not distorted as would inevitably happen if each part of the organism were drawn on a logarithmic scale. Therefore linear drawings are superimposed on a logarithmic grid. This is of course necessary, for otherwise the organisms would be unrecognizable.

KEY TO SYMBOLS

N = haploid
$2N$ = diploid
f = fertilization
m = meiosis

Acknowledgments. Whenever possible the drawings were made from live material and for this we are grateful to the help of Drs. R. D. Allen, R. F. Jones, and L. J. Kelland.

In all cases a large number of references were consulted, but those cited are the principal sources for each organism.

Bacillus sp.

Elongation and division in a rod-shaped bacterium.

Hartmanella astronyxis

The first two cycles show binary fission in this soil amoeba. The last cycle shows encystment, followed by a dormant period, and ending with germination. (After Ray and Hayes)

PLATE 1

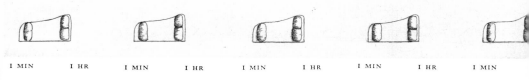

I MIN   I HR   I MIN   I HR   I MIN   I HR   I MIN   I HR   I MIN

PLATE 2

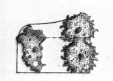 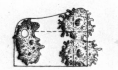 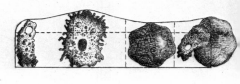

I MIN   I HR   I MIN   I HR   I MIN   I HR   I DAY   I MO   I YR

Chlamydomonas Reinhardi
The first two cycles show the asexual reproduction of the haploid individual. The last cycle shows the sexual cycle with fertilization, encystment of the zygote, meiosis, and liberation of the new haploid individuals.

Discorbis mediterranensis
The haploid growth phase of this foraminiferan is shown in the first cycle, and the diploid phase which includes fertilization and meiosis is shown in the second. These individuals are multinucleate, and growth of the shell is by the addition of increasingly larger chambers. (After Le Calvez)

PLATE 3

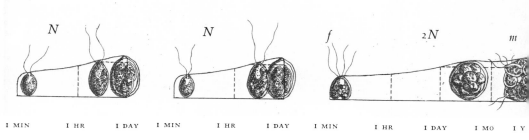

N N f $_2N$ m

| I MIN | I HR | I DAY | I MIN | I HR | I DAY | I MIN | I HR | I DAY | I MO | I Y |

PLATE 4

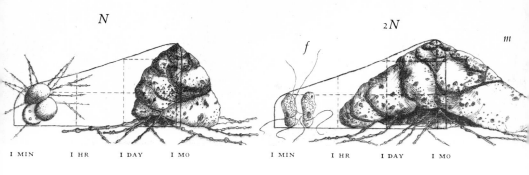

N $_2N$
                                  f m

| I MIN | I HR | I DAY | I MO | I MIN | I HR | I DAY | I MO |

Chondromyces crocatus

In this species of myxobacteria the hardened cysts contain many hundreds of rod-shaped bacteria. Upon germination the rods swarm by a gliding movement into larger and larger heaps; all the while they are feeding from the organic nutrients of the substratum. Finally the heap of bacterial cells becomes rounded, rises into the air, and as it does it exudes a noncellular gelatinous stalk. The cells migrate to the tips and become entrapped in hard resistant cysts.

Dictyostelium mucoroides

The spore of the cellular slime molds has a hard cellulose covering which splits upon germination, allowing a uninucleate amoeba to emerge. This amoeba feeds and undergoes numerous successive binary fissions (as in *Hartmanella*, Plate 2). When the food supply is depleted the amoebae aggregate to central collection points, and ultimately the cell mass builds a central stalk made up of dead vacuolate cells and an apical spore mass or sorus, in which each spore encloses one amoeba.

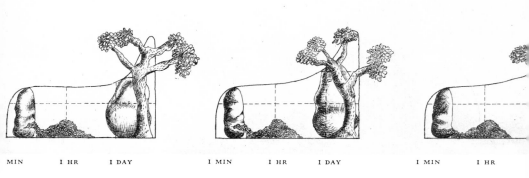

MIN   I HR   I DAY   I MIN   I HR   I DAY   I MIN   I HR

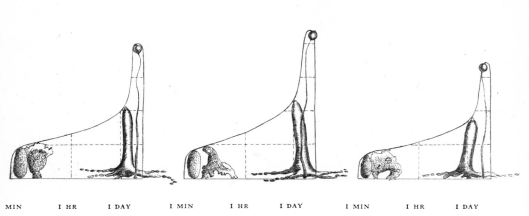

MIN   I HR   I DAY   I MIN   I HR   I DAY   I MIN   I HR   I DAY

Physarum polycephalum

In the myxomycetes the haploid spore germinates and the swarmer may interchangeably be flagellated or amoeboid. Following fertilization the multinucleate plasmodium grows, sometimes fusing with other plasmodia. Ultimately the large sprawling mass of feeding protoplasm will become condensed and nodular (as shown in the plate) and each nodule will give rise to a stalked fruiting body. Prior to the progressive cleavage and delineation of the spores in the sporangium, meiosis occurs. (After Howard, Yuasa, and Guttes)

Bryopsis plumosa

This marine siphonaceous alga has no crosswalls during its vegetative phases, and all the nuclei wander from one part to another by protoplasmic streaming. During gamete formation one of the branches seals off, meiosis and progressive cleavage ensues, and the gametes are liberated. Fertilization occurs free in the open water, and the zygote settles to the bottom where it ultimately becomes attached by means of a holdfast. (After Fritsch, Oltmanns, and Smith)

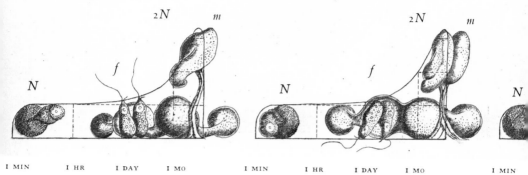

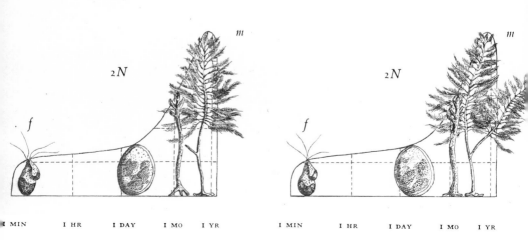

Volvox Rousseletii

In the asexual cycle of Volvox, which is shown in the first cycle, a large cell (or gonidium) in the posterior half of a mature colony will begin a series of divisions producing a daughter colony as a pocket inside the parent colony. The daughter colony will invert or turn inside out, and in the middle drawing of the first cycle one can see a colony just after inversion. The mature colony may repeat the cycle by producing more gonidia or it may produce gametes. Fertilization results in the formation of a resistant polyeder. Upon germination in the spring this will, following meiosis, form a small juvenile colony that will in turn give rise to successive asexual generations. (After Kelland and Pocock)

Ulva Lactuca

The sea lettuce is a common green alga along American shores. It has an alteration of haploid and diploid generations, and they are morphologically indistinguishable (except for cell and nuclear size). The zoospore or gametes (depending upon whether it is a haploid or diploid generation) are produced in the border zone of the mature plant. Note that in the successive stage of growth, *Ulva* goes through a *Ulothrix*-like stage, which is followed by a thickened filament (as in *Schizomeris*), and finally the blade flattens in two layers or sheets of parenchyma cells (see text, Fig. 4). There is no growth zone, but all the cells in the blade are capable of growth and division. (After Thuret)

PLATE 9

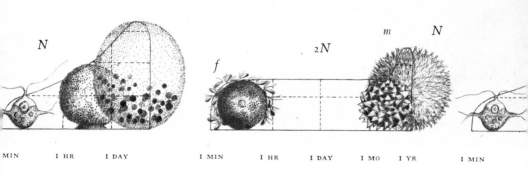

PLATE 10

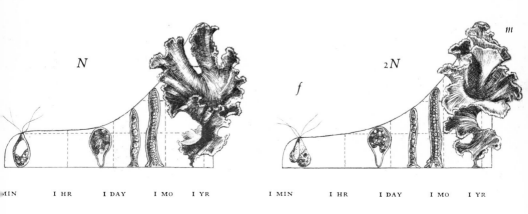

Laminaria flexicaulis

This large brown alga or kelp also has an alteration of generations, but in this case the haploid generation is small and inconspicuous (middle cycle) especially by comparison with the huge diploid generation (first cycle). In the big frond there is a distinct growth zone at the junction of the blades and the stipe, and considerable cell differentiation, especially conspicuous in the stipe where there are conducting systems made up of trumpet cells. (After Brown, Fritsch, and Sauvageau)

Fucus vesiculosus

The common rockweed is a brown alga in which the haploid generation has completely disappeared as a separate cycle, and the time interval between meiosis and fertilization is short. (After Brown, Fritsch, and Thuret)

PLATE II

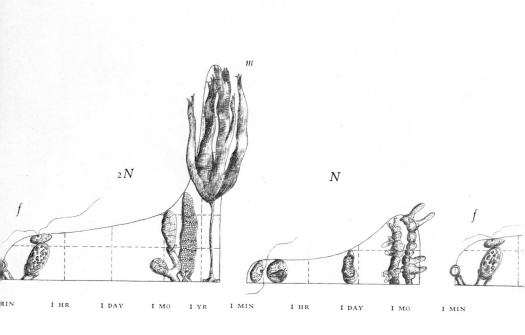

$_2N$

f

m

N

f

MIN   I HR   I DAY   I MO   I YR   I MIN   I HR   I DAY   I MO   I MIN

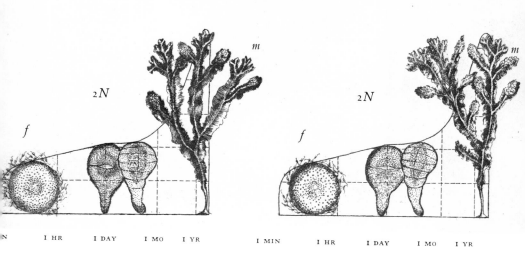

$_2N$

f

m

$_2N$

f

m

N   I HR   I DAY   I MO   I YR   I MIN   I HR   I DAY   I MO   I YR

Allomyces arbuscula
This water mold (aquatic phycomycete) has distinct haploid and diploid cycles, but they need not necessarily alternate. The diploid plant can produce thin-walled zoosporangia and thereby have repeated asexual generations. Only the production of resistant sporangia, which can withstand adverse conditions, will lead to meiosis and the production of the haploid sexual mycelium which contains the male and female gametangia. (After Emerson)

Coprinus sterquilinus
In mushrooms the hypha that emerges from the germinating spore is haploid and gives rise to the primary mycelium. Primary mycelia of compatible mating types will fuse and the nuclei of both will come together in pairs and remain in close association. This is the dikaryon condition found in the secondary mycelium and indicated as $N + N$ in the plate. Final nuclear fusion or karyogamy only occurs in the subterranean mycelium, and the protoplasm from this mycelium flows into the fruiting body in a matter of hours. The measurement of length, in the ordinate, is especially arbitrary in this instance. (After Buller)

PLATE 13

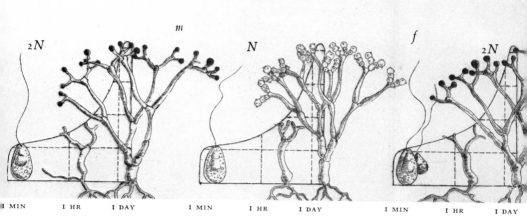

PLATE 14

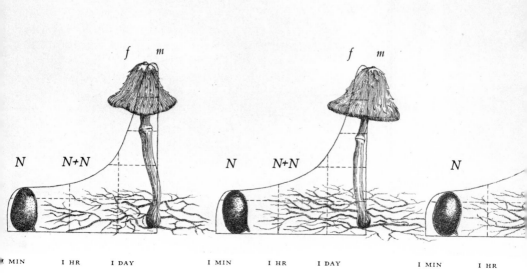

Polytrichum commune

The hair cap moss has an alteration of generations. The haploid generation (first cycle) begins with a delicate filamentous protonema on which buds develop into the main moss plant. At its tip the moss contains sexual organs (antheridia and archegonia), and after fertilization the diploid stalk rises up from the gametophyte supporting a spore capsule (second cycle). Meiosis occurs during the process of spore formation producing haploid spores. (After Bold, Brown, and Smith)

Dryopteris filix-mas

In ferns the haploid generation is small and relatively inconspicuous (first cycle). The spore germinates into a short filament which soon expands into the thin heart-shaped prothallus. Antheridia and archegonia are on the central region of the prothallus. Following fertilization the sporophyte burgeons upward, soon dwarfing the old prothallus (second cycle). The mature fern has haploid spores on its leaves. Meiosis takes place during the formation of these spores. (After Bold, Brown, Parihar, and Smith)

PLATE 15

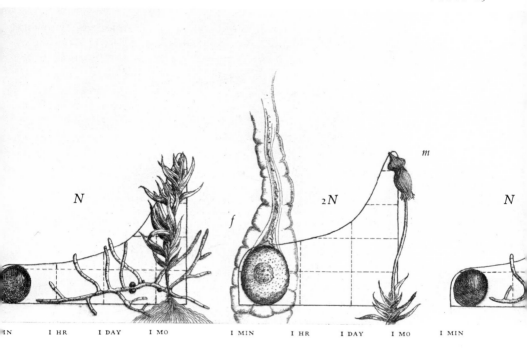

N   f   2N   m   N

I MIN   I HR   I DAY   I MO   I MIN   I HR   I DAY   I MO   I MIN

PLATE 16

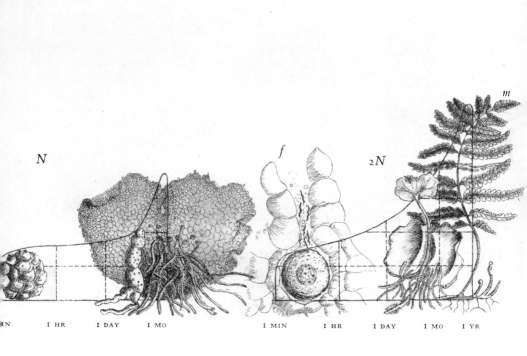

N   f   2N   m

I MIN   I HR   I DAY   I MO   I MIN   I HR   I DAY   I MO   I YR

Triticum aestivum

Wheat, like all higher plants, is characterized by having seeds. This is a resistant stopping place which interrupts the period of size increase. There is growth between fertilization and seed formation, a period of equilibrium, and subsequent growth upon seed germination. Wheat is an annual plant and therefore there is no cambium or secondary thickening. After the single spurt of one season the main body of the plant turns yellow and dies. (After Caruthers)

Sequoia gigantea

Sequoia is the largest tree and can be effectively compared to the annual plant shown above. Fertilization and the early growth to the seed stage are essentially similar, but because of the cambium and the possibility of secondary thickening, the size of the tree can increase enormously. As can be seen from Figure 1 in the text, the sequoia does not begin to set seed until it is sixty years old and eighty meters tall. (After Brown, Dallimore, and Jackson)

PLATE 17

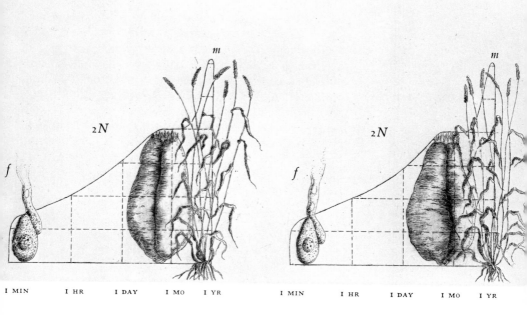

I MIN   I HR   I DAY   I MO   I YR   I MIN   I HR   I DAY   I MO   I YR

PLATE 18

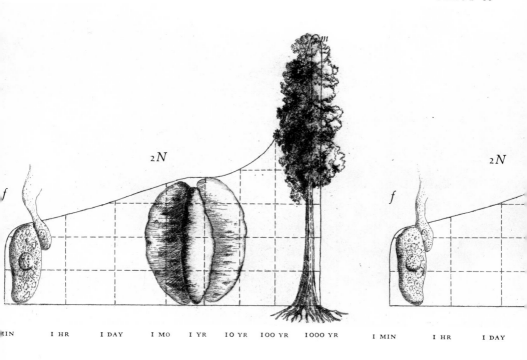

MIN   I HR   I DAY   I MO   I YR   IO YR   IOO YR   IOOO YR   I MIN   I HR   I DAY

MULTICELLULAR
ORGANISMS
(with soft motile cells)

Hydra viridis
In asexual reproduction the hydra forms a bud at the lower portion of the gastric column which eventually detaches (first cycle). A hydra may produce eggs and sperm (one large egg is shown on the body surface in the beginning of the second cycle) which give rise to a zygote in a resistant case. Upon germination a normal individual capable of budding is produced. This is not a set alternation of sexual and asexual generations, but as in *Volvox* (Plate 9) and *Allomyces* (Plate 13) and many other organisms, there is a series of asexual cycles when environmental conditions are constant and favorable, and sexual reproduction is apparently stimulated by the onset of adverse conditions. (After Brien, Reniers-Decoen, and Hyman)

Dugesia dorotocephala
Certain planarian flatworms can divide asexually by merely pinching in two (first and third cycle) or sexually (middle cycle). The gonads are within the worm, and the steps in the early embryology are elaborate and unusual. (After Child and Davidoff)

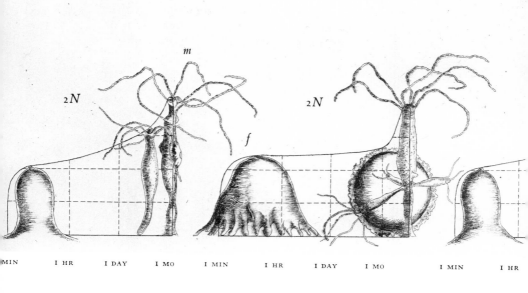

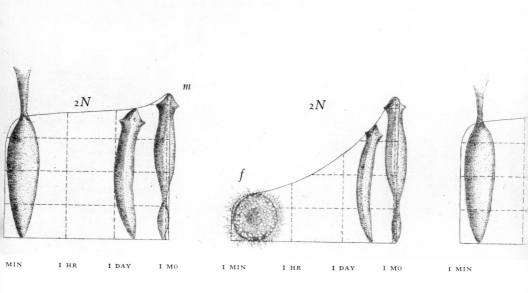

Melanoplus mexicanus
The grasshopper has an incomplete metamorphosis. The fertilized egg produces ultimately a miniature nymph, which can be seen just after emergence from the egg case (middle drawing in the first cycle), and after a series of nymphal molts the adult form is progressively approached. (After Chopard)

Danaus plexippus
In a monarch butterfly there is a complete metamorphosis from caterpillar to adult. In the pupa a great many of the tissues disintegrate, and the adult form is built up from the imaginal discs. It is for this reason that the upper curve of the cycle is shown to dip during pupation. (After Urquart)

PLATE 21

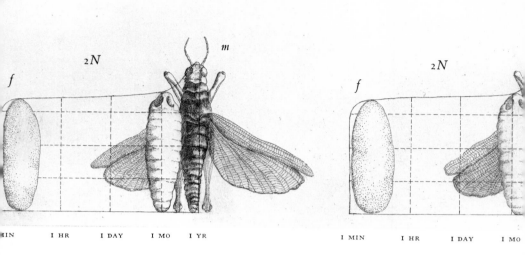

$_2N$ m

f

f $_2N$

I MIN   I HR   I DAY   I MO   I YR

I MIN   I HR   I DAY   I MO

PLATE 22

m

$_2N$

$_2N$

f

MIN   I HR   I DAY   I MO   I YR

I MIN   I HR

Arbacia punctulata
The sea urchin also has a dramatic metamorphosis. After fertilization the pluteus larva results, and eventually in a small region of the larva the bud of the adult begins. Therefore this is another case where metamorphosis is accompanied by a period of size decrease. (After Harvey)

Ciona intestinalis
In the tunicates there is also a metamorphosis between the tadpole larva (middle drawing of each cycle) and the sedentary adult. However in this case the decrease in size is relatively insignificant because it mainly involves the loss of the tail, for the head of the larva grows into the mature ascidian. (After Berrill)

PLATE 23

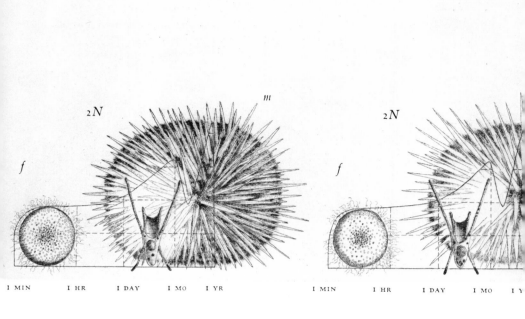

I MIN   I HR   I DAY   I MO   I YR   I MIN   I HR   I DAY   I MO   I Y

PLATE 24

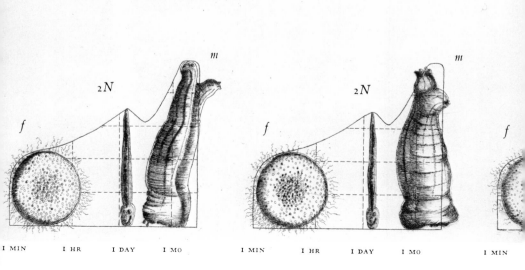

I MIN   I HR   I DAY   I MO   I MIN   I HR   I DAY   I MO   I MIN

Rana pipiens

As in the tunicates (Plate 24) the metamorphosis in the frog involves merely the loss of the tail and therefore is accompanied by a size reduction of minor importance. (After Taylor and Kollros)

Balaenoptera musculus

The blue whale is the largest animal. As can be seen from Figure 1 in the text, it first is capable of reproduction when it is five to six years of age and approximately twenty-two meters long. (After Slijper and Young)

PLATE 25

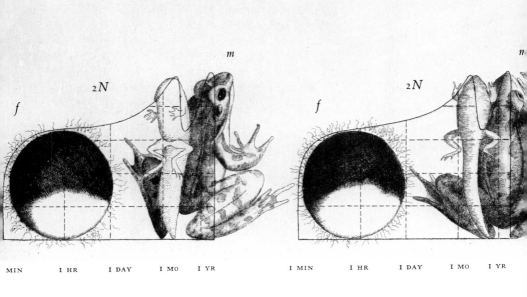

m

n

f $_2N$ *f* $_2N$

MIN   I HR   I DAY   I MO   I YR   I MIN   I HR   I DAY   I MO   I YR

PLATE 26

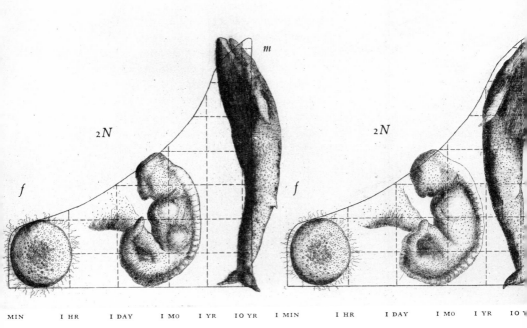

m

f $_2N$ *f* $_2N$

MIN   I HR   I DAY   I MO   I YR   IO YR   I MIN   I HR   I DAY   I MO   I YR   IO Y

Bougainvillia superciliaris

In the colonial hydroids there is both the cycle of the polyp (first cycle) and the cycle of the medusa (second cycle). In this particular species these two cycles are evenly balanced. The egg produces first a primary polyp which then branches into a large colony (first cycle). Certain specialized reproductive polyps transform into medusae which are liberated and free-swimming (second cycle). They contain the gonads and produce the egg and sperm. (After Allman and Berrill)

Agalma Pourtalesii

In the siphonophores, in contrast to the related colonial hydroids such as *Bougainvillia*, the individual persons, either polyp or medusoid in origin, take on many different functions and become appropriately differentiated. In the young form (middle drawing of each cycle) there are two budding sites: the upper one produces the float and the swimming bells while the lower one produces the cormidia which are groups of persons including feeding, protective, and reproductive individuals. (After Delage and Mayer)

PLATE 27

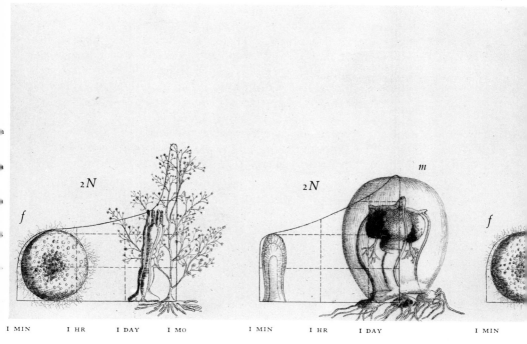

I MIN   I HR   I DAY   I MO   I MIN   I HR   I DAY   I MIN

PLATE 28

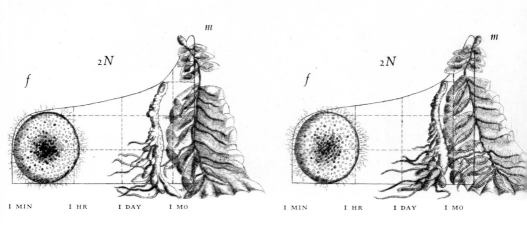

I MIN   I HR   I DAY   I MO   I MIN   I HR   I DAY   I MO

Cladonia cristatella

In the lichens there is a symbiotic association between a lichen and a fungus. The first cycle, which is illustrated here, is hypothetical because it is assumed that the fungus spore (first drawing) and the *Protococcus* alga (second drawing) unite to form the lichen. In point of fact this has not yet been observed but only surmised. It is known that lichens can reproduce by detaching groups of cells (sometimes call soredia) of both the fungus and the alga together (second cycle). (After Ahmadjian, Lamb, and Brown)

Camponotus ligniperdus

In the social insects the colony may begin by the individual development of a queen from an egg (first cycle). The queen, after fertilization, removes her wings, finds a small burrow, and begins to lay eggs. These neuter offspring take over nursing and foraging functions completely. Ultimately the queen, who as a rule is exceptionally long-lived, will give rise to new reproductives that can begin the whole process again. (After Wheeler)

PLATE 29

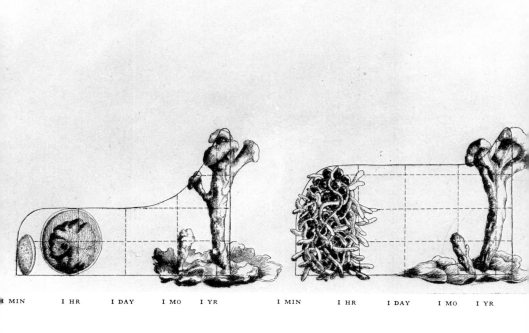

MIN   I HR   I DAY   I MO   I YR   I MIN   I HR   I DAY   I MO   I YR

PLATE 30

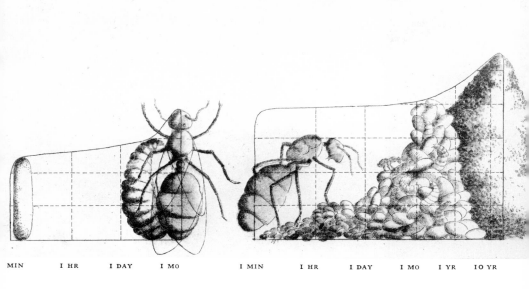

MIN   I HR   I DAY   I MO   I MIN   I HR   I DAY   I MO   I YR   IO YR

there is a loose association of individuals we refer to it as a society. Sometimes the distinction between society and population is unclear, for it depends upon the degree of closeness of the bonds that link individuals, but both are examples of the largest kind of biological unit.

We can superimpose upon this largely quantitative scale a qualitative one. This relates to the fact that the structure of all cells is not the same; in particular some are hard surfaced while others are soft. The bacterial cell, the plant filament, and the silicaceous diatom are examples where the cell membrane is closely associated with a rigid wall, while the amoeba and animal cells in general remain flexible and deformable. The reason for superimposing this structural dichotomy upon the size scale is that the type of multicellular organism produced depends both on the number of cells and the type of construction unit. From an engineering point of view one can do certain things with hard cells that are impossible with soft cells and *vice versa*; the consequences of this difference are profound. For instance, soft cells can move and wander in organized ways to specific locations, while hard cells, although immobile, can give a rigid skeletal frame to a wide variety of structures. As is so often the case with sharp categories, there are numerous intergrades between softness and hardness, but one cannot escape from the fact that the properties of the cell surface are of immense importance in the organization of the organism. Furthermore, like size, hardness is a property that can be seen and directly measured; it is not a matter for conjecture.

The period of size increase is one in which there are a variety of constructive processes occurring simultaneously. These processes are all to some extent interrelated and can be subdivided or artificially analyzed in a number of different ways. The system I would like to use here is one proposed previously (Bonner, 1952) that seems sufficiently straightforward to be useful. It consists of dividing the constructive process into three components: growth, morphogenetic movement, and differentiation.

By growth is meant an increase in living material; it is the synthesis of protoplasm. Of all the three categories this is the only one which has any direct bearing on the increase in size; in fact, it *is* size increase. But for various reasons it never, at least for very extended periods of increase, remains alone, and is always associated with differentiation.

Differentiation seems always to depend upon chemical differences in different parts of an organism and is always associated with division of labor. This is the basic reason why growth is ultimately accompanied by differentiation. Increase in size involves changes in the surface-volume ratios of any structure and this has profound consequences as far as the functioning of the organism is concerned. Metabolism itself is a function of all the tissue and therefore related to the volume (or the cube of the linear dimensions); the passage of food or fuel to the cells and the exchange of gases with the surroundings all are dependent upon the surface area (or the square of the linear dimensions). Weight varies with the volume since the specific gravity of protoplasm is roughly constant, while strength varies with the cross-section area. Therefore, as was mentioned previously, increase in size is accompanied with many structural changes that are virtually dictated by surface-volume relations. As a consequence growth and differentiation are hand-mates, for in the long run any significant growth will mean that the living machine cannot operate without some differentiation, some division of labor, some increase in complexity in our specific sense.

Morphogenetic movement does not by itself imply any synthesis or increase in size or any differentiation or regional change in chemical structure. It simply is the moving of parts from one region to another. As a rule this involves cells, although in some cases, such as oöplasmic segregation in certain eggs or fruiting in the plasmodial myxomycetes, there is a kind of morphogenetic movement which takes place within one cell membrane. Where cell movement occurs the key fact is that the cells can move, and that mobility is related to the properties of their surface. Hard-walled cells, such as those of many multicellular plants, are immobile and therefore their period of size increase is utterly devoid of morphogenetic movements. On the other hand multicellular animals with their soft, amoeba-like cells, undergo all sorts of morphogenetic movements, of which gastrulation is the most conspicuous. These movements are possible because the cells can move, and in most of the cases the cell movement is dependent on the soft walls or membranes permitting (and probably achieving) pseudopodial activity and amoeboid movement. It allows a growing mass of cells to redistribute its bulk in a fashion that will produce an effectively functioning individual. It is not solely dependent, as is the case with plants, upon the direction of growth to achieve a specific stage; it can move its cells about as well.

With these points in mind it is now possible to examine the period of size increase in different size groups, and for each group to consider the type of variation in the method of size increase, some of which will be imposed by the character of the cell surface.

Unicellular organisms

Bacteria are generally regarded as primitive unicellular organisms, and they are indeed small (Plate 1). They possess so many structural differences from ordinary cells (or what Picken (1960) calls eucells) that it has proved in many cases difficult and even dangerous to make comparisons between the two. The nucleus is certainly of a radically different structure, and the cytoplasm lacks any organelle equivalent to the mitochondria of eucells. In fact, from a metabolic point of view the surface of bacteria seems to perform enzymatic activities similar to those of the mitochondria of higher cells, and therefore it is conceivable that mitochondria are convoluted bits of surface that have passed into the interior of the cell as a result of the increase in size of eucells from those of bacteria.

It is known that in bacterial cells there is virtually a continuous synthesis taking place during all or during almost all of the life cycle. The separation phase is not marked or preceded by any significant halt in the synthesis of DNA. On the other hand, amoebae and other eucells show distinct phases in their cycle of active DNA synthesis, followed by extended periods of rest. (For a critical discussion see Mazia, 1961.) This difference applies only where bacterial cells are growing at a rapid rate; if they are barely growing, then they too will show pauses.

DNA is a substance of key interest and importance, but in itself it is not an accurate index of cell size. Dry weight is obviously more desirable, and it has been shown for both fission yeast (*Schizosaccharomyces*) and budding yeast (*Saccharomyces*) by Mitchison (1957, 1958) that there is a linear in-

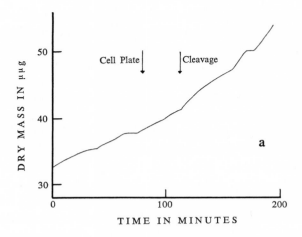

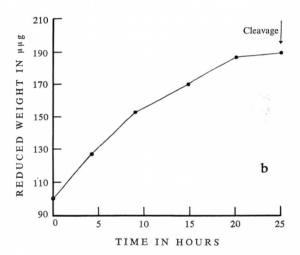

FIGURE 3. (a) Increase in the dry mass of a single cell (becoming two cells) of the fission yeast, *Schizosaccharomyces pombe*. (After Mitchison, 1957). (b) Increase of reduced weight of a single amoeba, *A. proteus*, between divisions. (After Prescott, 1955)

crease in the dry mass during the life cycle and that this is unaffected by the advent of cell division (Fig. 3a). This interesting fact was demonstrated by following individual cells and their progeny under the interference microscope. At the same time it was possible to follow volume changes, which did show a pause before division, but this pause was not reflected in the dry mass. In a more recent study Mitchison (1961) followed single bacterial cells (*Streptococcus faecalis*) and found the rate nonlinear, progressively falling off over the period between divisions. As he pointed out, dry weight is not equivalent to synthesis. In fact in these instances synthesis will decline progressively during the cell cycle, while the changes in the amino acid pool will keep the rate of dry weight increase constant.

Using a different technique, Prescott (1955) followed the reduced weight in a single individual of *Amoeba proteus*. He placed the amoeba in a cartesian micro-diver and followed the weight changes during the division cycle. In this case he also found that the rate of increase does not remain constant throughout the period between divisions but is progressively reduced (Fig. 3b).

There is a large amount of data on volume and length changes during the division cycle for other unicellular organisms, summarized by Adolph (1931), but the difficulty is to know how to compare this precisely with the more significant dry weight measurements. Nevertheless, provided suitable environmental conditions favorable for growth are maintained (and this has been assumed all along), there is either a linear increase in volume or length, or, more often, a progressive decrease as the moment of the next division approaches. In all cases there is some increase in size throughout the cycle, the only significant variable is whether the rate of synthesis remains constant or decreases. The only possible interruptions are produced by environmental causes such as lack of food or unfavorable temperatures. We are, of course, momentarily excluding the kind of pause in size increase that would be included under encystment and resting stages in general (Plates 2, 3).

The periods of size increase for different cell types vary greatly in their duration. But nevertheless there is one important correlation that may be made: in general bacterial cells have a much shorter generation time than single eucells, and even among eucells, smaller cells have a shorter time between divisions than larger ones (see Table 1).

In conclusion it may be said that unicellular organisms tend to increase in size continually, at least until they approach the moment of separation, of cell division. If there are any major pauses they do not appear to be inborn but the result of limiting environmental conditions. Size increase is the principal occupation of unicellular organisms when the environment provides suitable opportunities.

Aggregative organisms

There are two groups of organisms in which unicellular forms unite in a mass to produce a multicellular organism. These are the myxobacteria and the cellular slime molds. It is particularly interesting that this rather unusual behavior should be represented in two obviously unrelated groups and involving cell types of radically different construction: rod bacteria and small amoebae. It is undoubtedly a case of rather primitive convergence.

In the myxobacteria, using *Chondromyces crocatus* as an example, when the rods emerge from the germinating cysts they immediately swarm together (Plate 5). As they swarm they also feed, and each individual cell grows and repeatedly divides. Eventually the swarm will reach a critical size and then rise into the air to produce a small fruiting body bearing terminal cysts. It is not known if growth of the cells continues during the fruiting process, but since the cells are lifted away from the substratum into the air we may assume that they have little or no food supply, and therefore the rising into the air is largely a period of morphogenetic movement with little or no growth (see Quinlan and Raper, 1964 for a review of the development of myxobacteria). Thus there is first a period of growth of the individual cells (although the cells are always in contact and moving about in loose swarms) followed by a morphogenetic movement which ends in dormant cyst formation.

The cellular slime molds show an even more dramatic cleavage between a unicellular growth phase and multicellular morphogenetic movement and differentiation (Plate 6). (For reviews see Raper, 1940, Shaffer, 1962, Bonner, 1959a, 1963.) In this case, upon emergence from the spore case, the amoeba feeds and divides and after each division the daughter cells repel each other (Samuel, 1961). Feeding eventually stops, usually by consumption of the food supply, and after a pause of some hours the amoebae stream together to central collection points (aggregation). The resulting cell mass proceeds to differentiate into spore and stalk cells, and as a result of some interesting morphogenetic movements a slender stalk is raised into the air supporting (at least in many species) a single spherical spore mass or sorus. While it is known that cell divisions do occur in the cell mass, it is presumed that this does not represent growth or increase in size, for the cells have been without food for some hours. Instead it must be regarded as size reduction by cell division, using the food reserves built up from the previous period of feeding as the source of energy.

As was pointed out many years ago by Harper (1926), this is almost a perfect case of a natural separation between the growth phases and the other developmental phases in the life history. Growth, in this case of unicellular amoebae, occurs first and stops. This is followed by a period of morphogenetic movement and differentiation which produces a multicellular individual.

However, from our point of view these aggregative organisms are truly intermediate between unicellular and multicellular forms. As we have defined it, a life cycle involves size increase and separation; and therefore during the vegetative stage the amoebae undergo a series of true unicellular life cycles, each generation of which has its own size increase or growth phase. At aggregation there is a massive collection of amoebae by morphogenetic movement, and therefore from the point of view of the multicellular part of the life cycle, size increase occurs very suddenly by the chemotactic accumulation of cells to central collection points. Following aggregation, the remaining developmental processes take place without increase in dry weight, although there is an increase in length by cell redistribution. This is not an uncommon occurrence in periods of development, and various examples will be examined when we later consider periods of size equilibrium in the life cycle. It should be added that while size increase is very rapid in the cellular slime molds, it is far slower in the myxobacteria in which swarming and cell growth go on simultaneously.

Because aggregative organisms are intermediate between the uni- and multicellular condition, they serve to point up a number of problems that are of significance to the whole concept of life cycles and levels or units of organization. In the case of the unicellular stage while the cells are not physically attached, they do communicate with one another; first by repelling one another and later by attracting one another across considerable distances (up to 500μ) to central collection points. Perhaps the most significant feature of cells actually touching each other in a multicellular form is that by this close contact communication is readily effected. But vegetative and pre-aggregation amoebae do communicate with one another, although undoubtedly the extent of the communication is limited by the great distances between the cells. We can now conceive of a third and more primitive condition where single cells have no relation or communication at all among themselves, and then it is obvious that between the extremes there must be every kind of intergrade.

However, a population of single cells of any microorganism usually will have some kind of communication, although it may be of a very rudimentary sort. A group of flagellates in an aquarium will, both by their metabolic products and their competition for food, manage to affect the distribution and the growth rate of their neighbors. They will, even more, be at the mercy of environmental conditions: the size of the aquarium, the temperature, the food supply, the presence or absence of predators, etc., yet nevertheless to a minor extent their successes may be affected by information (materials) they pass among themselves. If cells are literally touching, then these stabilizing mechanisms can be and are very much more effective. In terms of cell communication the difference between unicellular populations and the multicellular organisms is mainly a matter of degree.

Multinucleate organisms

A consideration of multinucleate organisms will be more by way of parenthesis than a detailed treatment. The lack of proper formation of cross walls or cell membranes following nuclear division or multiplication is a phenomenon found in all the principal groups of organisms. Therefore rather than being a major method of construction, it is more likely an aberrant phenomenon due to a lack of wall formation that occurs sporadically throughout the animal and plant kingdoms, and especially among protists. There are a few cases of special interest in their own right which will be considered, but the examples among higher forms show no significant novel features from their related multicellular forms and therefore will not be considered separately.

In the case of bacteria and yeasts it is possible by the alteration in temperature and a number of chemical agents, as Nickerson and others have shown (review: Nickerson and Bartnicki-Garcia, 1964), to suppress cell-wall formation and produce multinucleate filamentous forms. This is perhaps not so surprising in the case of yeasts since presumably they stemmed from filamentous ancestors. Again in both these cases the phenomenon is a lack of something, rather than a positive approach to constructive problems.

The situation in ciliates is highly specialized and very different. The macronucleus contains the genome repeated many times, and therefore a *Paramecium*, for instance, is effectively a multinucleate individual. During the period of growth following cell division there is both an increase in the cycloplasm as well as in the macronucleus.

The micronuclei, which are diploid, are only concerned with conjugation and autogamy. The other characteristic of a ciliate is its fantastically complex surface structures. The lacework of superficial small delicate structures has no equal in other forms, and in each cell cycle these structures enlarge and duplicate themselves. The amount of new cortical structures that are built up is limited, and even these for the most part seem to arise from pre-existing surface structures. No ciliate completely remakes its surface each cell or life cycle, but it adds and alters to the pre-existing half it inherited from its mother, so that when its full size is achieved it is complete. The period of size increase in ciliates resembles regeneration, for in each cycle there is an elaborate restoration process following the surgery of binary fission.

Myxomycetes provide another interesting example of multinucleate forms (Plate 7). In this case there is a long period of size increase of the multinucleate plasmodium. As with the cellular slime molds and the myxobacteria, when the food supply is exhausted the protoplasm will stream into concentrations (which amount to plasmodial aggregations), and a series of small fruiting bodies will arise as the result of often quite elaborate streaming morphogenetic movements. At the very end of development (following meiosis) there is a cutting out of the nuclei by progressive cleavage to form spores. In this life cycle it is the period of size increase which is conspicuously multinucleate, and one must presume that this structural condition is in some way correlated with the locomotory and feeding mechanism.

As a final example of multinucleate organisms the coenocytic algae (and numerous fungi) should be mentioned. Many forms, such as *Bryopsis* (Plate 8) and *Caulerpa* are, during their vegetative period of size increase, entirely without cross walls. Again progressive cleavage only occurs just prior to gamete or spore production. There is the particularly interesting case of *Acetabularia*, where the many nuclei remain in one mass (as in the macronucleus of ciliates) during the whole growth phase, and in this case gamete formation is preceded first by the breaking up of the large nucleus into many small ones, followed by progressive cleavage.

Multicellular organisms

The most important type of size increase is, of course, achieved by multicellular organisms, and we assume that in most cases multinucleate forms had true multicellular ones as their ancestors. There is, however, the opposite possibility (for which there are some vociferous proponents) that multicellular forms arose from multinucleate ones by a progressive cleavage. It is even quite reasonable that both phenomena occurred; but since it is impossible ever to test any of these possibilities, it is perhaps more fruitful to point out that both cases involve a nuclear multiplication with the nuclei remaining physically attached to one another. That there may or may not be cross walls is perhaps not of such critical significance, for communication between nuclei can be readily achieved either way. The only really radically different method of becoming multicellular is that found in aggregative organisms. In most cases one starts with a single nucleus and all the cells in the growth mass are derived from it, while in aggregative organisms one may have the coming together of genetically diverse nuclei. There are also intermediate cases which have both systems: for example, heterokaryon formation in fungi where the individual filaments grow clonally, but they can fuse and thereby achieve an aggregative status.

It is, of course, possible to grow a cellular slime mold from one amoeba, and often the characteristics of a clone of unicellular organisms have been compared to the growth of a multicellular form, the only difference being that in one case the cells stick together and in the other not. But this has an important implication. In unicellular forms the cell cycle fulfills our requirement of growth and separation and therefore constitutes a life cycle. But if the daughter cells stick together in one larger organism, then the cell cycles become submerged in the new, larger unit. The result is that we have jumped from one order of magnitude to another in our life cycles, and the reason for the jump is entirely due to the fact that in one case the cells separate and in the other they do not. This distinction also applies to aggregative and multinucleate organisms, for in both cases there is a period where the nuclei and associated cytoplasm are contiguous. The units we have accumulated thus far are the nucleus and its surrounding cytoplasm, which form a cell cycle, and the life cycle which may consist of a cell cycle or numerous cell cycles in contiguous groups of cells. In shifting from one level to the next the period of size increase is more extended and therefore the period of size decrease (or separation) is more dramatic.

If we were here to consider in detail the period of size increase for all organisms our task would almost be without end. Instead I should like to examine some general trends in size increase. The first and most useful distinction is between hard- and soft-walled forms. As has already been stressed, this produces major differences in structure which are of considerable significance.

The filament is a basic hard-walled unit. It is a stiff cylinder whose walls are lined with fibrils of cellulose or chitin. Because of the stiffness, shape is determined primarily by the direction of growth. For instance, in the case of a single filament the growth is in a straight line. This growth direction is reflected in the direction of the mitotic spindle and by the perpendicular orientation of the cell plate. These ideas were emphasized by Julius Sachs, who pointed out that the spindle was always (or almost always) along the long axis of the cell and the cleavage plane across it. This is known to botanists as Sachs' rule, and Hertwig showed the same thing to be true for the soft cells of animals.

If one were to vary the direction of growth (or it could be expressed in terms of the direction of the spindle axis), it would be possible to produce a whole variety of shapes over and above a filament. These possible permutations have been examined in more detail elsewhere (Bonner, 1952); here we shall give a brief summary (Fig. 4).

If occasional cells grow at right angles to a straight line, the result will be a branching filament (e.g. *Stigeoclonium*). This can occur either in three dimensions or in two on a flat surface. In the latter case if the branches tend to fill all the available space the result will be a flat disc (e.g. *Coleochaete*). If a single cell divides alternately at right angles it will produce a flat rectangular sheet of cells (e.g. *Merismopedia*). If it divides successively in the *x*, *y*, and *z* axes it will produce a cube (e.g. *Eucapsis*). If a number of filament cells divide at right angles to the main axis and subsequently divide again in what now amounts to a tangential plane, there will be a progressive thickening to form a true thallus (e.g. *Shizomeris*). With further divisions this may form a hollow cylinder (e.g. *Enteromorpha*) which with further size increase may collapse into two sheets of cells (e.g. *Ulva*, Plate 10). In these larger forms the filamentous nature of the cells is progressively lost and replaced by a parenchyma. In all the largest algae the parts become thickened in all directions by parenchyma-like cells to form a solid structure. This condition, of course, is found up to and including the higher plants.

Before commenting further on this main trend of plant size increase, another type of filamentous size increase should be mentioned. This is the accumulation of many relatively loose filaments in fasces, a condition that exists among a few algae and many fungi. There is good reason to assume it arose among the fungi independently a number of times; it is found among relatively simple groups such as the *Aspergilli* and *Penicillia* in the form of rudimentary coremia, in many fleshy ascomycetes such as the cup fungi, and finally in the basidiomycetes such as mush-

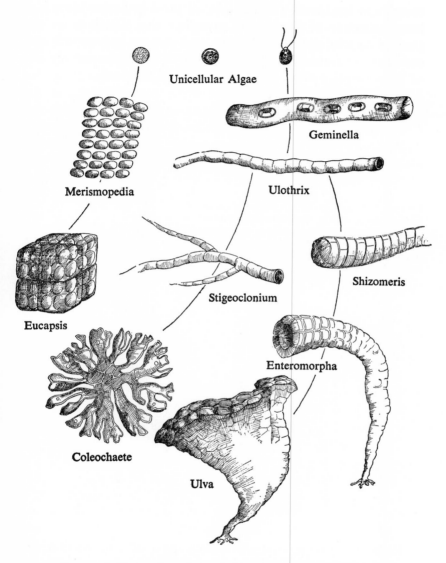

Unicellular Algae

Geminella

Merismopedia

Ulothrix

Eucapsis

Shizomeris

Stigeoclonium

Enteromorpha

Coleochaete

Ulva

FIGURE 4. The various shapes achieved among the algae can be interpreted in terms of growth direction (or cleavage patterns) in organisms with rigid cell walls. By changing the direction, the frequency, and the distribution of cell elongation and division, a wide variety of shapes is possible.

rooms and toadstools (Fig. 5). In all these instances there is a loose collection of filaments or mycelium that somehow are organized so that the over-all shape and pattern has smooth and consistent contours. The whole period of size increase consists of a cooperative effort of many individual filaments.

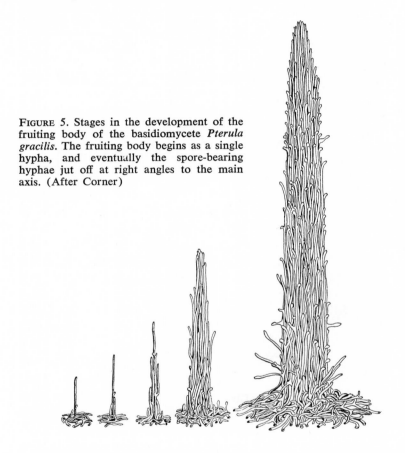

FIGURE 5. Stages in the development of the fruiting body of the basidiomycete *Pterula gracilis*. The fruiting body begins as a single hypha, and eventually the spore-bearing hyphae jut off at right angles to the main axis. (After Corner)

Returning to the parenchymatous type of construction, we see the different kinds of features that evolved to cope with the problem of increased size. Of these by far the most significant is meristematic growth. In *Ulva* (Plate 10) even though a large blade is produced, any cell anywhere on the thallus is capable of division. This means that when a cell in the center expands, all the walls around it must expand by plastic deformation. The result is that the walls themselves can never become completely rigid. If greater strength is needed, the development of a growth zone permits older regions to remain firm and rigid while only the actively growing parts are pliable. By having the meristem at the apex, a firm stem or a blade may be continually added to, or the meristem may be intercalary and placed between two firm parts.

As size increased, another problem arose in the evolution of plants: namely their support on land. Some of the largest algae, the marine browns, such as *Laminaria* (Plate 11), *Macrocystis*, and other kelp, do have meristematic growth, but they are supported by the water and in fact could not stand erect in air. Increased size in air has required, besides a meristem, very strong skeletal fibers for support, such as we find in wood.

In the evolution of the ferns there is an increase in size, and we are indebted to Bower (1930) for a detailed analysis of the relation of size to the structure of the supporting tissue. He showed that not only is the total amount of vascular tissue increased in larger ferns and other vascular plants, but its distribution and pattern is far more complex (Fig. 6). He attributed this to surface-volume relations and the fact that strength varies with the cross-section area and weight with the volume. In large gymnosperms and angiosperms the development of secondary growth in the form of a lateral meristem or cambium that completely surrounds the trunk is even more effective in permitting a gigantic pillar of wood slowly to accumulate.

Besides the differentiation involved in the elaboration of the meristem, there are many other differentiations of higher plants associated with size increase. There is not only the supporting function of the vascular tissue, but the conductive function as well. Roots buried deep below the ground must receive the products of photosynthesis, and leaves far up in the air must receive water and salts from the soil. Beside these major divisions of labor associated with size, there are many minor ones.

In animals with their soft cells, the construction problems are very different. The most conspicuous difference is that during embryonic development there is a considerable amount of cell movement which can rearrange the cells after cleavage. The direction of cleavage is therefore in general not so critical; however, this statement is far more applicable to some organisms than to others. It is conspicuous that among animals there are two types of development: mosaic and regulative (or determinate and indeterminate). In the former the cleavage pattern is fixed, and any experimental interference with it will result in mon-

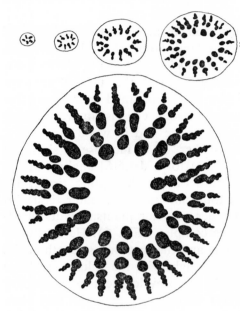

FIGURE 6. The distribution of vascular tissue in cross sections of different sized roots of the monocot *Colocasia adorata*. They are all drawn to the same scale (× 29) and note the increase in complexity of the bundles with increase size. (After Wardlaw and Bower)

strosities, while in the latter the ability to accommodate to alterations is striking. Both have gastrulation and other morphogenetic movements; the difference lies largely in their ability to replace parts and restore themselves after a disturbance. We will consider these matters further, but here let us merely emphasize the fact that the difference does not truly represent a fundamental divergence in the method of development. Rather, mosaic forms have become rigid and inflexible in their steps, while regulative forms can easily reverse and change. But the basic steps must be similar, the chief difference being the degree of rigidity or irreversibility. The same kind of distinction is found in the ability of some animals to regenerate parts while others cannot.

If we look at the gamut of animals from the smallest to the largest, it is surprising how basically similar is their construction.

In all cases that arise from a fertilized egg (and this includes the vast majority) there is first a period of cleavage followed by an intense period of morphogenetic movement. It is only subsequent to the gastrulation process that any real differentiation or division of labor begins. The mechanism of gastrulation differs widely in various groups, but in some cases the amount of cell mixing involved is considerable. What function this serves is unfortunately not known, although one might suppose that cleavage has given rise to a variable population of cells and that gastrulation sorts these cells out in such a way that the pattern of the future organism is laid down. Certainly such a sorting does take place in cellular slime molds after cleavage (Bonner, 1959b; Takeuchi, 1963), but we are anticipating our discussion of the analysis of the steps during the period of size increase.

After the so-called regionalization has been established, differentiation and growth go hand in hand to form the major structures of the body. In comparison to plants, animals are extremely active and mobile, and therefore the division of labor for even a small animal is considerable, but as size increases the degree of elaboration becomes formidable. When one considers the digestive, the circulatory, the endocrine, the nervous, and the respiratory systems and all their elaborate structure, the complexity is impressive. Moreover, many animals have made the shift from water to land, and the supporting skeleton and muscular attachments have had to make the appropriate accommodations. In some cases the skeleton is outside and in others it is inside, and even though both bring with them their own engineering problems, the basic result is the same.

Skeletal structures are the only really rigid parts of an animal. All the soft parts may grow and expand without difficulty; there is no equivalent to plant meristems. The one striking exception is in some sessile organisms such as colonial hydroids, which not only look like plants but grow meristematically as well (Plate 27). However, in animals the endoskeleton is exceedingly hard and accretionary growth is necessary. For instance, in the junction between the diaphysis and the epiphysis of the long bones of mammals there is a meristematic zone responsible for bone elongation. Animals with exoskeletons are sometimes as fortunate, for example the gastropod who continually deposits his shell; but some are less so, like the lobster who must periodically shed and rebuild his entire skeleton.

To summarize in very general terms for soft-celled animals, one can say that the steps are cleavage, then gastrulation, followed by a period of the combined activities of growth, differentiation, and further morphogenetic movements. The extent of the post-gastrulation period, of course, varies with size. The only part of the growth that need be meristematic is that associated with certain types of skeletal tissue. By contrast plants never have morphogenetic movement, and all, not just part of their form, is achieved by directional and regional growth. In larger plants all the growth is meristematic. The difference between these two major groups can entirely be ascribed to differences in the nature of the cell wall.

There is another interesting feature of the period of size increase in both animals and plants. Frequently, and this arises irregularly in different groups of all organisms, there is a radical shift in the whole structure of the organism. The most obvious case is metamorphosis in animals. After a long growing period a caterpillar will turn into a butterfly (Plate 22), a pluteus larva will turn into a sea urchin (Plate 23), a tadpole will turn into a frog (Plate 25). The list could be extended endlessly with familiar and exotic examples, all of which would tell the same story. In each case there is a shift in the type of construction of the growing animal, in some cases more radical than in others. It should also be noted parenthetically that sometimes at metamorphosis there is a decrease in size before subsequent growth (e.g. sea urchins) and in other cases there is a long pause at the transition period (e.g. diapause in insects).

Differences in form between adult and larva might well have adaptive value and hence are not entirely reducible to the problems of constructing an adult. Thus it is not surprising that many parasites have different structural phases depending upon their relation to the host. The most extreme examples appear where there is more than one host, and the parasite has a special form for each one of its host phases. For instance the liver fluke (*Fasciola hepatica*) produces cercaria in pond snails and adult flukes in the livers of sheep.

The very same kind of polymorphism is found in plant parasites. For instance the wheat rust (*Puccinia graminis*) has wheat and barberry as hosts in successive stages of its life cycle. The sporulating structures and the spores themselves are different for each phase, and therefore the transition from wheat to barberry can be considered the equivalent to animal metamorphosis. In plants, however, there is another interesting difference. Frequently the forms will be the haploid and diploid phases of the life cycle. Animals never have any development of a gametophyte generation, and all their permutations in form during the period of size increase occur during the diploid stage (with the exception of a few permanently haploid animals). Furthermore in animals, even though there may be a momentary size reduction during metamorphosis (e.g. echinoderms), on the whole there is an over-all increase in size from egg to adult. In plants generally both the diploid and haploid generation start from scratch. For instance in a fern the small, delicate prothallus develops from a spore, but eventually it succeeds in producing sperm and egg which start the larger sporophyte generation (Plate 16). An exception would be the case of the moss, for the simple reason that after antheridia and archegonia have formed at the tip of the large gametophyte, the fertilized egg remains *in situ* with the result that the gametophyte and the sporophyte remain attached (Plate 15). By this accident the result is that the over-all size of the moss plant steadily increases in size.

The main point, however, is that during the period of increase for both hard- and soft-walled organisms there is a possibility of shifting in a variety of interesting ways. Each shift is an integral part of the life cycle, and it shows that the period of size increase in the life cycle may be either smooth and without significant events or it can be full of momentary stops and reverses, particularly of different phases of size increase each with its own distinctive character.

Colonies of multicellular organisms

In making the transition from uni- to multicellular organisms we moved from one level to another. The life cycle of a multicellular organism contains numerous cell cycles, while a unicellular organism only has one. In considering multicellular colonies, populations, and societies, we now make one further step, for here we have what amounts to a collection of life cycles of multicellular organisms that are somehow grouped together. This occurs in a variety of different ways, and for each it is important to know the means of integration or communication and also the general method by which the group increases in size.

In general we think of the individuals as being separate from each other in any kind of social grouping, but there are certain types of colonial organisms consisting of a collection of multicellular individuals that are physically attached. A primitive example would be a mushroom or even a simple bread mold such as *Mucor* or *Neurospora*, for in these cases there is a continuous mycelium and from it sprout many fruiting bodies which may, depending upon the species, be simple branching filaments (e.g. *Mucor, Penicillium*) or massive aggregations of filaments (e.g. a mushroom). Among higher plants there are many that will either have a long rhizome that continually sends off shoots as it travels through the ground (e.g. Solomon's seal) or will have leaves or branches that touch the ground, root, and start a new but attached individual plant (e.g. a walking fern). Among animals this type of compound colony is found in a number of groups. The sponges are frequently compounded together, although as a rule their structure, especially in the leuconoid sponges, is so irregular that it is hard if not impossible to determine where one individual ends and the other begins. Many of the most striking cases occur among the coelenterates. The colonial hydroids are a collection of separate "persons" all attached by a common plumbing system (Plate 27), but the pinnacle of this type of colony is to be found in the special group of siphonophores (Plate 28). Here not only are the individuals attached but each individual has become highly specialized and there is a marked division of labor. In fact some of the persons, such as the pneumatophores and the swimming bells, have almost lost their identity as separate individuals; it would seem as though the siphonophores have evolved backwards and gone from a true collection of individuals back to a single highly differentiated multicellular individual. The implication is that there has been an increase in the amount of communication between parts, and as a result instead of becoming larger the colony has become more highly integrated. Two further groups that exhibit colony formation are the bryozoans and the ascidians. In both cases there are numerous species that form compound groups of individuals. In brief, this is a common phenomenon among plants and animals.

The most usual method of size increase in these colonies is by continual outward growth, with new individuals simply sprouting like buds from a branch. In higher plants this may literally happen by new buds appearing on roots, stems, or even leaves. In ascidians and hydroids, except that the tissue that grows outward is of a very different nature, the new individuals arise in much the same fashion. As with a single multicellular organism, the new individual arises by failure of the growing parts to separate. It is interesting that in fresh water hydra the ability to separate was regained, so that while hydra's ancestors are colonial, in hydra itself the individual persons separate after they have formed by budding (Plate 19). The failure of growing buds to separate is the most direct way to achieve colonies of multicellular organisms.

It is not, however, the only way. There are a number of cases where fusion of separate aggregated individuals takes place. Certainly, as Buller (1933) discussed in detail, this is the case among many fungi, for the hyphae themselves will anastomose. Numerous examples are found among the basidiomycetes, where a network of mycelia may fuse to form a group of separate but joined fruiting bodies. The same possibility exists among sponges, and in colonial hydroids the anastomosis of stolons is dramatically similar. Not only will the stolons fuse so that the gastro-vascular canal becomes continuous, but also the new amalgamated network will in concert build new polyps.

In all the cases of colonies of multicellular organisms, however, the principal method of size increase of the whole colony seems to be by expansive growth. It may be that there is a period of aggregation or amalgamation, but almost without exception this is preceded and followed by expansive growth.

Another characteristic of the growth, however, is of special interest. If the colony is continuously enlarging, it is also generally true that the individuals (polyps or fruiting bodies) will remain roughly the same size and all increase is put into making more individuals. Thus we may assume that the persons (of a coelenterate or a fungus) have not lost their identity or their control mechanisms, any more than have separate individuals in a population.

The fact that they are physically connected may or may not be of great significance for the communication between individuals. In some cases, such as those of higher plants, the main purpose of the connecting link between individuals seems to be that of propagation. Once that is achieved the presence or absence of the connecting link is of little importance. In fungi and in colonial hydroids food can be passed from one region to another of the large network of the colony. There are also among hydroids nerve connections between parts which integrate the whole colony, and furthermore Crowell (1953) showed some extremely interesting cycles of hydranth regression and replacement in *Obelia* and *Campanularia*, indicating that the whole colony is performing certain actions as a unit.

Perhaps the best indication of the degree of communication between the individuals of a colony is the extent that the whole colony has a specific form and a division of labor. The coelenterates are particularly interesting in this respect because within one group they have run the gamut from loose networks to tightly organized, bilaterally symmetrical colonies, such as sea pens. In the highly developed siphonophores the division of labor has reached such a point that the individuals have lost many of their properties and become highly specialized. One might assume that the reason that siphonophores appear to revert to the type of organization characteristic of an ordinary multicellular organism is that the communication between parts is so good that they can differentiate as easily and in the same way as a single multicellular organism.

Earlier we characterized a life cycle as having both growth and separation. In colonies of multicellular organisms the separation is clearly lacking, just as the separation of cells was lacking in the beginning of the formation of multicellular organisms. Therefore the life cycle now encompasses the growth, the duplication, and the separation of the whole colony; it is the colony which is the new cycling unit.

Populations and societies

Attached colonies of multicellular organisms are often large, but by far the largest grouping is that of a population. The distinction between a population and a society is a difficult one to draw, for it is a matter of degree. In a society the integration of the members is closely knit; there is extensive communication between the close (but physically separated) individuals. In populations the individuals still form a loose unit, but the extent of intercommunication of parts is relatively limited.

In both, as with colonies of multicellular organisms, it is not just the individuals that are associated, but there is an alignment of their whole life cycles. In colonies the timing of the cycles of the individuals is under fairly rigid control: the new individuals keep budding in succession, or as in the siphonophores, all the individuals arise simultaneously so that the cycle of the colony coincides with the cycles of the individuals. In some societies the relation of the timing of the individual life cycles to that of the whole cycle is also clearly defined, but in loose populations the relation is often vague and the individual cycles overlap in a haphazard fashion. However there may be some regularity to the individual cycles, for the majority of animals and plants are geared to environmental changes and their reproductive periods are affected by seasons.

The organisms that are least equipped for any kind of social integration are plants, because communication between individuals is greatly restricted. Tactile, visual, olfactory (i.e. chemical), and sound signals are the principal means of communication for animals, and to send and receive these kinds of signals there are special receptors and often special emitters. In plants the only method is chemical: the production and diffusion of substances and their direct effect on neighboring plants.

As a result plant populations are poorly integrated. They grow largely as a result of slowly spreading over an area, ultimately to be checked in their expansion by the size of the habitable area, the competition with other species for space, and the available nutrients. Once they have achieved this level of growth they will go into a period of equilibrium during which individual life cycles will rise and fall, constantly overlapping each other.

There are numerous animal populations that are hardly any more integrated than a population of plants, but because animals are mobile we more often find that during certain specific seasons the population is more closely knit, more social, than at others. For instance many birds and mammals congregate only during the reproduction or nesting period, while others, such as starlings, congregate during the non-reproductive periods of the year. In each case there may be a mixture of ages among the group, and therefore their life cycles overlap to some extent although the timing of the reproductive period is determined by annual cycles in climate and is the same for all.

In loose populations what integration does exist may be used to avoid any cycling of the whole population. A strong case could be made for the notion that from an adaptive point of view one of the most important functions of a population is to keep the number of individuals from becoming too large or too small, for either may result in disastrous destruction. Many ecologists in the past have examined this point, and recently Wynne-Edwards (1962) has suggested that communication systems between individuals might provide mechanisms which control the number of individuals. Cycles are well known and documented in detail for many populations, but even these great fluctuations in numbers lie between definite upper and lower limits. Homeostasis of the populations is their most significant feature as far as their adaptive maintenance and survival is concerned.

There are many good examples of more integrated societies among mammals and birds, but with respect to the absence of cycling and the maintenance of a steady state of their numbers, they are no different from looser populations. In the most extreme case of social grouping of animals, however, the situation is radically different.

Social insects have apparently arisen independently a number of times, and despite great differences in detail, there are certain common features (Wheeler, 1922). In the first place all insect societies are basically a single family and revolve around a king (whose days may be few) and a queen who remains permanently with her gigantic brood. In the event of her death there are other mechanisms for secondary reproductives, or the production of new kings and queens to take over. If this occurs the colony is maintained in equilibrium and there is no new growth period.

New colonies may start either from a solitary fertilized queen or a king and a queen may literally begin their family from scratch (Plate 30). The first brood of sterile offspring are workers to share in the toil of nest building, larva feeding, and all the other duties of the society. As the eggs keep coming the community becomes larger and larger, the sole source of individuals being the queen mother. Ultimately the size may reach an equilibrium where the worker deaths equal the new arrivals, and as long as there is a queen this equilibrium can be maintained.

In some cases, such as the honey bees and army ants, there is a division of the workers among rival queens and therefore the new colony will start with a considerable initial labor force. The difference between these two, by analogy with the cellular level, is between an individual arising from a spore or a fertilized egg, as compared to an individual arising from a multicellular individual splitting in two (as in some worms, e.g. Plate 20).

But the most remarkable aspect of insect societies is the fact that the workers divide the labor. Not only that, they even become modified in shape according to their function in the colony. Again increased size has been accompanied with a division of labor; differentiation is the ordering of the necessary complexity to keep large and highly integrated groups in operation.

It is interesting that in these closely knit insect societies the life cycle of the individual and the life cycle of the group are one and the same. For basically it is the life cycle of the queen upon which the whole society is built. It is true that this generalization has exceptions, since secondary reproductions may take over in some cases when the queen dies, or there may be a continuity of queen generations maintained by groups of workers that weather the transition and help start the new cycle. But basically insect societies are fantastically large families with only the parents ordinarily serving as the germ plasm to carry through to the next cycle. As was so forcefully emphasized by Wheeler (1911), the fact that the workers are sterile gives them a status analogous to the soma. They cannot have life cycles of their own, because although they increase in size, and are well coordinated multicellular individuals, they cannot duplicate and separate; only the queen can do that. Therefore these most integrated societies are unique in that they themselves are life cycles and not simply a series of life cycles fused and overlapping. In this they resemble colonies of multicellular organisms where again the most integrated cases, for example, the siphonophores, also show a single cycle closely resembling that of a single multicellular organism. This clinging to a single straightforward life cycle may have certain advantages which could be understood in terms that we have already suggested: The life cycle is the unit of evolutionary change and through it innovations and eliminations are possible. By keeping the colony or society within the span of one life cycle, the changes can be controlled more immediately and directly.

Conclusion

The basic principles that I have tried to illustrate in this discussion of size increase can be set down as a series of statements:

The larger the organism the longer the period of size increase and the more complex and highly differentiated is the result.

Size increase is generally accompanied by an increase in the amount of intercommunication of parts; some units (cells or multicellular organisms) are more closely knit than others and with increased unity there is a more effective intercommunication.

The rigidity of the cell walls or membranes largely determine the type of construction during the period of size increase. If the walls are rigid the cells are immobile, and the direction of growth is the principal means of governing shape. In populations of plants in which the multicellular individuals are also non-motile, the complexity of relations between parts is minimal. If the walls are soft and the cells motile, there is invariably a period of cell movement or rearrangement during the period of size increase. Usually multicellular animals are also motile and may (although not necessarily) have complex social relations.

The units of construction in a series of increasing magnitude are cells, populations of separate cells, multicellular masses (and multinucleate ones), attached multicellular colonies, societies of organisms, and populations of multicellular organisms (Fig. 7). In unicellular organisms the life cycle is the cell cycle; in multicellular organisms it is a larger span covering numerous cell cycles (the larger the organism the greater the number of cell cycles).

In populations and societies the multicellular individuals still have their own life cycle, and except for seasonal or externally induced mating seasons the individual life cycles tend to overlap within the population. Populations do not have rigid cycles; rather they can more effectively be described as maintaining a constant

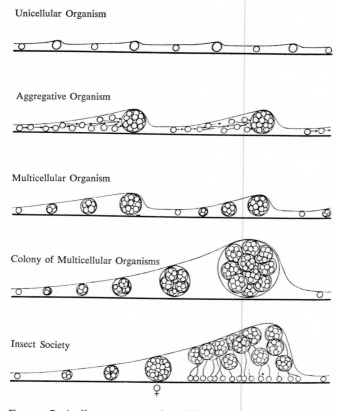

FIGURE 7. A diagram comparing different kinds of life cycles, both simple and compound. The single small circles represent cells, and a group of small circles enclosed in a larger one represents a multicellular organism.

number of individuals, and if there are any cyclic fluctuations, these are within upper and lower limits.

However, in colonies of multicellular individuals and insect societies, the life cycle is that of the colony or society as a whole. The life cycle, therefore, may or may not correspond with that of the larger group, but in either case it remains the basic unit of natural selection.

THE PERIOD OF SIZE DECREASE

We are faced with the difficulty that writing about biological phenomena forces one to put the facts in linear order, while a true picture would show them simultaneously branching in all directions in three dimensions. For instance it is artificial and unsatisfactory to separate the life cycle into periods of size increase and size decrease, for this tends to do just what I am specifically trying to avoid, namely chopping up and undermining the idea of the importance of the life cycle as a continuous process. However I am not clever enough to discuss everything at once, so this bit of dissection and analysis, although imperfect, is unavoidable.

As has already been emphasized, a life cycle consists of a period of waxing and one of waning, and the latter is often connected with the requirement of separation which is an integral part of the cycle. If an organism splits in two it both separates and becomes reduced in size simultaneously. Therefore one kind of size diminution is characterized by some sort of abrupt cleavage or fragmentation.

There is also another kind of size diminution which is gradual and is not in any way connected with the separation part of the life cycle. Principally this is the process of senescence, when there may be a slow shrinkage in one part of the cycle. Each of these two principal categories of size decrease will be considered separately, and each will be considered at all size levels: unicellular, multicellular, etc.

In the case of abrupt size decreases it is possible to make a further subdivision of some significance. In some of the instances the whole individual fragments so that each part of the whole becomes a new seed, a new beginning for another cycle. By contrast, there are those cases where the whole individual remains intact and merely fragments off a small part of its substance to start a new generation, a new cycle. This process may be performed repeatedly.

Total and abrupt size decrease

The most obvious kind of total and abrupt size decrease is that of binary fission in unicellular organisms: the bacterium, or the amoeba that splits in two, both halves separating (Plates 1, 2). In the majority of these cases the halves are roughly equal, although in some cases there is a certain amount of haphazard inequality. Budding yeasts, for instance, seem to inflate the daughter cell slowly, while the parent cell remains roughly the same size. However in the closely related fission yeasts a cell plate is thrown half-way across the mother cell and each daughter increases in size.

Cleavage in two is not confined to single-celled organisms, but it is also known among some invertebrates. Certain worms, such as some species of flatworms like *Dugesia* (Plate 20) and various annelids, may split in two at the equator (see Berrill, 1961). In all these instances the division is definitely not equal, but after a period of size increase both portions will be similar in size and construction. In some social insects each new cycle is started by a solitary queen spawning her new family from the beginning, but in others the new queen takes some of the workers with her while some remain with the old. The result is a splitting of the whole colony, a character found, for instance, in honey bees and army ants. Schneirla and Brown's (1950) observations on this splitting in army ants is particularly pertinent. They observed that the old and the new queens remained together for a considerable period, all the while competing for the attention of the workers. Each would become surrounded by a group of attentive slaves, and progressively the two groups would tend to separate and repel one another, apparently by some sort of odor difference. Ultimately the two groups would cleave totally.

Besides cleavage in two, there are many cases where the organism will break up into numerous small units. This is also a case of total size reduction, for all the parts of the organism will be involved in this fragmentation.

Many unicellular forms possess this ability. Usually it appears to be a case of an extended period of size increase followed by a spasm of repeated, successive divisions. This is found among some flagellates (Plate 3) and numerous ciliates, although the latter are more appropriately considered multinucleate forms.

Sometimes the repeated subdivision is asexual in nature, and sometimes sexual. In the latter case meiosis or reduction division is involved, and this of course means two divisions. Therefore inevitably the production of gametes results in at least the parent cell fragmenting into four daughter cells. This is precisely what occurs, for instance, in the flagellates *Chlamydomonas* (Plate 3). Another example would be yeast, where one may find either four or eight haploid ascospores.

Some aggregative organisms have total size reduction and others not. In myxobacteria all the cells (except for a few that are presumably trapped) produce the stalk and then either turn into microcysts (e.g. *Myxococcus*) or into large cysts (e.g. *Chondromyces*, Plate 5). Among the slime molds *Acytostelium* is unique in that, as Raper and Quinlan (1958) showed, all the cells produce the delicate cellulose stalk, and then all the cells turn into spores, ready to be scattered to disseminate the stock elsewhere.

Multinucleate organisms, of which ciliates were already mentioned, frequently separate by total cleavage. This is in part a reflection of their multinucleate nature. They produce a large body with many nuclei, and then suddenly this is rapidly cut up into cells (usually uninucleate) by what is known as progressive cleavage. It is as though the processes of cell division and nuclear division had been separated; nuclear division occurs gradually over a long period of time while cell division occurs in a sudden final flush. In many of these cases the cleavage is preceded by meiosis. This is the case for myxomycetes (Plate 7) and gamete production in various coenocytic algae (e.g. *Acetabularia*, the sexual phase of *Hydrodictyon*, etc.), while in others (such as the asexual phase of *Hydrodictyon*) the progressive cleavage is not preceded by meiosis and the spores are asexual.

Among multicellular organisms the total fragmentation of the whole body into more than two parts is a relatively rare phenomenon. To give a few examples, a filamentous alga may be literally broken into fragments, each one of which will start a fresh filament; the same is true for the mycelium of some fungi, the thallus of lichens, and the breaking up of rhizomes of some perennial plants.

Among higher animals it is almost an unknown phenomenon,

largely because we shall consider budding off a parent stock (such as hydra) as a case of partial rather than total splitting of the multicellular organism. After all, in this case the parent remains stable and keeps giving off buds or sprouts, which is quite different from the whole organism's splitting into three or more relatively equal parts.

There is one rather perfect case: the formation of quadruplets in armadillos, where a single egg cleaves and at a later stage separates to produce four identical multicellular embryos. The process is consistent from generation to generation; it is a fixed and inherited part of the life cycle.

Partial abrupt size decrease

By partial size reduction is meant that some of the organism is cut off to produce a new generation while the remainder continues essentially without alteration. This is in contrast to total size reduction, where all the parts to some extent require restoration. The step from total to partial is a significant one in that it implies a division of labor; only part of the organism is aimed towards reproduction, while the remainder remains stable. The stable portion may even repeatedly give off the reproductive bodies over a considerable span of time.

This particular division of labor is, of course, Weismann's soma and germ plasm. In some invertebrates and in most plants it is now known that the potentiality to produce a new generation is not merely restricted to a special group of cells, but rather any cell may, if given the opportunity, perform this function. In plants there is some evidence that it is possible to propagate a whole new carrot plant from one root callus cell (Steward, et al., 1961). The point is that even though the germ cells may normally arise in a restricted area, this does not mean that these cells are necessarily unique in being able to propagate another cycle.

Therefore instead of thinking of potentialities, it is more useful simply to designate those cells that are germinal and those that are not. The straightforward position we are taking here is that not all the cells are used in propagation; there is a division of labor and only some separate (hence the abrupt size reduction) and start a new life cycle.

In unicellular organisms such division of labor is hard to find. It is conceivable that a budding yeast could be interpreted in this way, for the old cell remains fixed while the new one grows off it as a bud. The only difficulty with such an interpretation is that before final separation the cells are roughly the same size, which brings the case rather closer to cleavage in two, as mentioned earlier. Also the fact that yeasts are presumably derived from filamentous ascomycetes gives them a rather special status as unicellular organisms.

Among multinucleate forms there are some genuine cases of considerable interest. For instance among the suctorians, which are no doubt derived from the ciliates, there is a budding off of a motile embryo from the sessile adult. The formation of the buds occurs in numerous interesting ways, as Guilcher (1951) has emphasized: there may be one or more buds carved out on the adult surface, and in each one a part of the macronucleus enters. Eventually it detaches, sometimes by turning almost completely inside out, to swim about as a ciliated larva with organized rows of cilia. After a while it will settle on some surface, lose the cilia, and take on the adult suctorian morphology.

Again we find that there may be sexual and asexual structures produced by multinucleate individuals. In this case, however, not all the substance of the organism is used up in the manufacture of the spores or gametes. In the green alga *Bryopsis* (Plate 8), for instance, certain of the petals or branches of their continuous tube are closed off, and progressive cleavage as well as meiosis takes place. By sealing off one branch, a gametangium has been produced. Much the same type of phenomenon may be found in the terminal hyphae of various phycomycetes, but in this case the result may be either sexual or asexual. *Allomyces* (Plate 13) or other members of the aquatic phycomycetes produce terminal zoosporangia and also have terminal antheridia and oögonia. Among the terrestrial phycomycetes, *Rhizopus* and other Mucorales produce asexual sporangia and progametes which form zygospores when two oppositely sexed hyphae meet and fuse. It is interesting that in some of these cases the spores may also have more than one nucleus, but they still are small in size. In all these cases of separation of reproductive bodies from multinucleate filaments, it must be remembered that the bulk of the filament is not involved, but only small portions of it. Furthermore it can repeatedly cut off such reproductive bodies.

Among aggregative organisms certain of the cellular slime molds have a distinctive division of labor between reproductive cells and soma, but the character of the soma deserves special note. In *Dictyostelium* the posterior cells become spores and the anterior cells become stalk and are incapable of propagation, as was demonstrated in detail by Wittingham and Raper (1960). It is further indicated from the withered appearance of their nuclei (Bonner, 1959a) that they are in fact dead cells. They have become so specialized as supporting tissue that once they have built their cellulose skeleton they expire. It may, at first glance, seem rather absurd that a division of labor should be rewarded by cell death, but then it must be remembered that supporting cells of many plants die (i.e. the wood fibers of a tree) and furthermore those slime molds with dead cells in their stalk produce much longer stalks than those without any cells (*Acytostelium*). Presumably this is because the dead cells with their cross walls are extremely helpful in supplying supporting cross-struts or trusses. For their particular function, being alive would be an unnecessary and perhaps even an inconvenient attribute.

If we turn now to multicellular forms, a major distinction can be made between those organisms which bud off single cells and those which give off multicellular buds. It is not possible to divide these two categories sharply into sexual and asexual respectively, because although it is true that all gamete formation is a matter of the liberation of single cells, there are also many types of asexual reproduction that involve single cells also.

Multicellular asexual buds are found in many groups of plants and invertebrates. For instance the lichens, presumably to keep the algal and the fungal partners together, produce a small tangled knot of fungus hyphae and algae filaments (a soredium) on the upper surface of the thallus (Plate 29). Liverworts have small cups with multicellular lozenge-shaped gemmae that become detached and start a new plant. Among higher plants the powers of regeneration are sometimes so extensive that they can produce the equivalent of a multicellular bud. For instance in the extreme case of the celebrated *Bryophyllum* plant, a leaf falling on the ground can produce not one but numerous new plants from new buds formed at the notches in the leaves.

Among animals, sponges have asexual gemmules. These are particularly remarkable in that instead of a special region of the sponge being cut off as a gemmule, the cells that make up the gemmule apparently aggregate to central collection points. Not only that, but other cells keep wandering in to provide the original cell aggregation with rich food reserves, and then to cap everything, a group of amoebae carry in the spicules which have been built elsewhere in the sponge and bring them to the surface of the gemmule and cement them in with military precision.

The coelenterates have more conventional budding methods. Hydra has a localized region of high growth activity which produces a miniature individual (Plate 19). Eventually (and this is the key point) the bud becomes detached, although Brien and Reniers-Decoen (1952) found an interesting somatic mutant that failed to do so, at least in the normal manner, and showed some of the characteristics of its colonial ancestors (Fig. 8). In the larval stages of some species of medusae there is also a budding off of new individuals. Sometimes the larva (scyphistoma) produces another individual like itself; in other cases there may be a huge structure that produces new individuals repeatedly (strobila), like a stack of paper cups in which the uppermost one becomes detached in succession. In all these examples of coelenterates one important point must be underlined. In each of these reductions in size by budding, this is not the only such period during the life cycle; in fact, all the examples given also have a sexual phase with single-celled gametes. Fertilization produces a larval form, which in the case of the hydroids is the polyp stage. By budding, this stage produces a medusa which contains the egg and sperm (Plate 27). It is said of those forms which lack the medusa stage entirely (such as hydra) that the medusa has degenerated as a stage, and as a result the gametes are produced directly on the polyp. The evidence for this contention is that there are among hydroids some forms in which medusae show some structure but remain abortive and attached to the polyp. By contrast the Scyphozoa have emphasized the medusa stage and show a great contraction of the embryonic period. The general significance of these shifts in emphasis are a matter for later

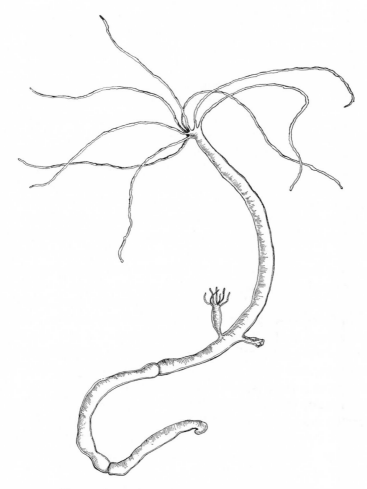

FIGURE 8. An individual hydra (*H. viridis*) in which the ability to degenerate and regress at the foot end is lost. Here two stolons that have become partially walled off are evident. (After Brien and Reniers-Decoen)

discussion; here we are only concerned with the fact that these organisms have two periods of size reduction.

As has already been stated, the same is true for some higher invertebrates that possess a dramatic metamorphosis. In echinoderms such as sea urchins the adult stage begins as a small wart on the pluteus larva, and in insects with complete metamorphosis the adult arises from the imaginal discs which comprises only a small fraction of the entire larva (Plates 22, 23). In both these cases all the larval tissues are lost except to provide organic materials as a source of energy for the radical new changes. This differs from budding or strobilization in the coelenterates in that it is not a means of propagating new individuals but is a tremendous alteration of the size increase activities of one individual by a sudden size reduction in mid-stream.

If we turn next to the consideration of multicellular organisms that propagate by giving off single cells, it is clear that the extent of the size reduction is automatically greater. It is the rule in multicellular forms (although there are a few exceptions) that the gametes or the asexual spores are uninucleate. This means that all the information for future propagation and expansion lies within this one cell; in effect this is the maximum amount of size reduction that can occur, for no structure less than a nucleus and its surrounding cytoplasm has this capacity.

Single-celled asexual spores are found among some algae, but they are particularly common among all the groups of fungi. Here we are really considering those fungi that are not coenocytic but truly multicellular, although often the distinction between these two states is not sharply recognized by fungi. They also exist in mosses and ferns that, as do many of the fungi, have an alternation of spore-bearing (sporophyte) and gamete-bearing (gametophyte) structures (Plates 15, 16). In our terms this is another case of two size-reduction periods in a life cycle, accompanied, as has already been discussed in the previous chapter, with two periods of size increase.

Among animals there are no asexual spores, although some forms (e.g. aphids, cladocerans, rotifers) will reproduce parthenogenetically, and functionally this amounts to single-celled spore formation. This, however, is a modification of the sexual process and not something developed over and above the sex mechanism.

Gamete formation is found among all groups of multicellular plants and animals. Since most cells are roughly of the same order of magnitude, and since the process of meiosis is approximately the same for all sexual organisms, the method of size reduction and the actual reduced size are more or less standard.

The variation comes from the type of structure that produces the gametes, whether it is hard-walled or soft, sessile or mobile, large or small. In fact the larger the organism the greater the amount of abrupt size reduction following gametogenesis. The extremes are those of the giant sequoia or the blue whale; in each case the life cycle, in one swift separation of gametes, has jumped from an object hundreds of tons in size to an object so small that it can hardly be seen with the naked eye. These are indeed the largest and the most abrupt size changes that occur in life cycles.

Social insects may also have an abrupt size decrease; since the insect society is a family, all that is needed is a male and a female to give rise to the whole mass of neuter offspring. A termite colony, for instance, will undergo size reduction by budding off fertile males and females. These sprout wings and can easily separate from the colony; thus the size is reduced from the thousands of individuals in a whole colony to two reproductives on their nuptial flight. The newlyweds leave together and start a new colony. The analogy between these reproductives and the germ plasm remains compelling.

Gradual decreases in size

None of the gradual decreases in size that we shall consider here is in any way connected with separation, as were the majority of the abrupt decreases previously discussed. Instead they are usually associated with senescence and decay. However it must be said that senescence may not necessarily be accompanied with size decrease, and size decrease may not mean decay; the two often but not invariably go together.

The only way that size decrease can be achieved in single-celled organisms is by starvation, and this is possible only in soft-walled forms because a hard wall prevents deflation (although not a decrease in dry weight). It is certainly true that if amoebae are starved they slowly become smaller. This obvious fact is relatively insignificant because it is not a built-in method of size reduction but is a simple direct effect caused by adverse environmental circumstances.

In multicellular forms there are some invertebrates, in particular some species of sponges, coelenterates, and ascidians, that have a standard device for size reduction under adverse conditions. They form so-called reduction bodies, which are collections of cells surrounded by some sort of membrane. They appear to be a last ditch reserve by which some cells are gathered in a safe place to await more favorable times. But again these steps are a direct result of the environment; the sponge does not form these bodies under normal growing conditions.

There is, on the other hand, one well-established case of inborn size decrease. As previously mentioned, it was discovered by Crowell (1953) that in the hydroids *Campanularia* and *Obelia* there is a regular cycle of growth and regression that occurs in waves throughout the cycle. Any particular hydranth will grow, recede, and grow again with considerable regularity. Since the second individual formed at the same place is really a totally new one, this reduction can properly be considered to be a significant phase of the life cycle of an individual polyp. This interesting example is unusual among coelenterates or even hydroids. In fact some of the organisms that are thought to be virtually without aging processes are the related sea-anemones.

Gradual size decrease has also been demonstrated for mammals. Not only is there a reduction of the size of supporting tissue which may cause a decrease in stature, but also there is a demonstrable decrease in the number of cells. This, as Strehler (1962) points out in his recent review, has been demonstrated for the brain of a bee, the human cerebral cortex, and possibly in muscle cells as well. But in all these cases the decrease is very slight, and in the human usually in the post-reproductive period. Size decrease due to senescence is quantitatively insignificant in higher animals and is not involved in the reproductive part of the cycle. In higher plants, so far as I am aware, there are no good examples of inborn or controlled gradual size decrease in the life cycle.

Conclusion

The period of size decrease may be abrupt or gradual. In the latter case, the significance of this slow decrease is largely unimportant in the functioning of the life cycle, with a few interesting exceptions, such as in the case of the hydroid *Campanularia*.

On the other hand, abrupt size decreases are key points in the cycle. They may either be at metamorphosis, when there is a radical change in the form of the individual from the larval to the adult stage (e.g. the insect arising from the imaginal discs of the larva) or they are at the point in the cycle when one individual separates into two or more.

In the latter case the separation may be of such a nature that the two or more parts are all reduced in size and need to go through a restorative period of size increase, as is true with most unicellular forms after cell division. In multicellular forms there is characteristically a division of labor between those cells that separate and those that do not. In some instances these separation bodies are sexual in nature (i.e. germ plasm and soma) while in other cases there is an asexual bud. In the latter case the bud may be unicellular (e.g. a spore) or multicellular (e.g. a true bud), but in the production of sexual bodies the gametes are invariably unicellular because of the exigencies of meiosis. However, from the point of view of size changes, the fact that some of the reductions are sexual and some asexual is of no significance; these designations are used here merely as helpful means of identification. The presence or absence of sexuality is solely a concern of the control of the amount of variation, a matter to be considered later, and has no direct bearing on matters of size change during the life cycle.

THE PERIOD OF SIZE EQUILIBRIUM

There are many different types of pauses in life cycles and they can be classified in a variety of ways. Since we have taken size as our principal means of subdividing the cycle, it again may be useful to separate the periods of equilibrium during the different size phases of the cycle. We can first examine equilibrium when the cycle is at its minimum, that is, when the organism is at its smallest size; then we may examine periods of rest or pauses during the phase of size increase; and finally we may look at equilibrium stages when the organism is at its maximum size within the cycle. The examples we shall discuss come from widely divergent groups of animals and plants; apparently periods of equilibrium have arisen time and time again during the course of evolution. We shall also see examples of two kinds of equilibrium: static and dynamic (steady state).

Equilibrium at the point of minimum size

Among plants a hard-walled resistant spore is exceedingly common. It is found among bacteria, many algae, and almost all fungi, bryophytes, pteridophytes, gymnosperms, and angiosperms. In many of these the spores are asexual, such as the spores of bacteria, aplanospores of algae, conidia and sporangiospores in fungi, and the spores of the sporophyte generation of mosses and ferns. In most of these the resistant body is uninucleate, although in some, such as the sporangiospores of terrestrial phycomycetes, there may be more than one nucleus. In a few, such as the gemmae of liverworts, the "spore" is multicellular.

Dormant asexual spores are almost unknown in animals. However some of the unicellular soft-walled forms, such as *Hartmanella* and related soil amoebae, have resistant cysts (Plate 2). The multicellular gemmules of sponges have an equilibrium period, but aside from this there are among multicellular animals no good cases of asexual dormant stages at the point of minimum size. In fact, it should be pointed out that in *Hartmanella* it is not even entirely clear that the cyst is the minimum point in all senses, for it becomes small partly through loss of water and not dry weight.

It is possible to put stationary sexual spores in two categories: those that are gametes and those that are zygotes. Examples of the latter are found among the algae, such as the resistant polyeders of *Volvox* (Plate 9) or various members of the Chlorococcales (e.g. *Hydrodictyon* or *Pediastrum*), and the zygospores of terrestrial phycomycetes such as *Mucor* or *Rhizopus*, the hardshelled oöspore of aquatic phycomycetes such as *Achlya* or *Saprolegnia*, the ascospores of ascomycetes, and the basidospores of basidiomycetes. In some of these meiosis takes place before the hard wall sets (e.g. ascospores), and in others meiosis takes place upon germination (e.g. zygospores).

The resistant dormant zygote is absent in many other lower plants, especially among the algae and certainly among all the higher plants. The matter of seeds in angiosperms will be discussed presently, for they are neither the immediate zygotes nor are they the point of minimum size in the life cycle.

Among animals fertilized eggs are frequently hard-shelled and resistant and in any event lie dormant. This is the case for numerous invertebrates (e.g. myriads of insects), fish and reptiles. In birds, even though there is a hard egg case, there is no appreciable resting period. In fish and reptiles an egg may remain inactive for some period of time. It almost appears that the lower one proceeds down the scale, the longer the period of dormancy may be extended.

There are some unusual and interesting examples among mammals. For instance, in the case of Alaskan fur seals and numerous species of bats there is a regular delayed implantation of the zygote. This period of inactivity apparently serves as an adaptation to time the birth of the offspring correctly; in the case of the seal, birth must occur during the short summer season on land, and the bat has a period of winter hibernation that intervenes between fertilization and parturition.

Resistant gametes which can remain viable for extended periods are also common. Characteristically it is the male gamete that has this attribute, and this occurs where, because of a terrestrial habit, the sperm has lost its means of aquatic locomotion. Resistance to adverse conditions and general ability to remain viable over long spans of time are apparently a substitute for locomotion, or at least directed locomotion. If pollen, which for locomotion uses the wings of insects or the vagaries of the wind, had a means of flying directly to the pistil, then perhaps the ability to remain dormant would not be so ingrained. There are no equivalent situations among animals, although the terrestrial fungi, such as a number of ascomycetes including *Neurospora*, produce spermatia (or microconidia) that are capable of fertilizing the female ascogonium. In this case they can also produce a new mycelium by parthenogenesis, a property in common with many eggs.

There is something uninspiring about such a list of examples. Obviously it is of much greater interest to know the conditions for the initiation of dormancy and the reasons for its stoppage, and especially its adaptive significance in relation to the general nature of the particular organism. But these matters must wait, for here we are concerned solely with a description of the different kinds of pauses in the cycle.

Periods of size equilibrium which interrupt the period of size increase

Periodic stopping of size increase, particularly among lower forms, may be due to simple external causes such as lack of food or unfavorable temperatures. We are not especially concerned here with such superficial examples but rather are looking for those instances where there is a regular and predictable stoppage.

A possible exception might be those cases where the unfavorable condition produces a special structure designed specifically for a pause. For instance, in myxomycetes drying conditions will cause the plasmodium to round up and seal off in a hard case. This so-called sclerotium can exist in a resting state for long periods of time, only emerging under suitable conditions of humidity and temperature. It amounts to a sort of lower-plant hibernation.

A good example of a pause occurring early in the growth phase is that of the seeds of angiosperms. After fertilization there is an extremely active period of growth of the embryo as well as the surrounding tissue so that by the time the seed is formed the embryo is already made up of many cells, the main shoot and root axes are established, and masses of food are stored in the cotyledons. Although the extent of seed longevity has been greatly exaggerated, there are known cases of seeds being successfully stored over three hundred years. This means that for this great period the embryo remained in what amounts to suspended animation.

Another classic example is diapause in insects. This occurs in a relatively later period of development, but again there is an absolute stoppage that will frequently extend throughout the whole winter season. As a rule, after the cocoon and chrysalis is formed, the pupa, following considerable internal modification, reaches an equilibrium point and remains essentially unchanged for long periods of time. Upon reactivation, further violent internal modifications are set in motion and eventually the totally altered adult emerges.

Besides insect diapause, a number of other invertebrates have a stopping point in mid-growth. This is true, for instance, of a number of parasites. The nematode *Trichina* penetrates the flesh of its host at a larval stage, and once the small worms are inside the muscle they encyst and remain dormant. In fact, they can be released only when the flesh is eaten and digested; then they germinate and continue to grow further. In the previous chapter it was pointed out that metamorphosis sometimes involves a size decrease, and here we are adding that often there is an extended period of no appreciable change in size, or size equilibrium.

One final example is the annual growth (or rather the annual period of no growth) in higher plants, especially deciduous plants in northern climates. This phenomenon is closely related to season, although even in the tropics many trees shed all their leaves at one time and go through at least a short period without growth. If the leaves are shed, no photosynthesis is possible; and without any intake of energy growth must come to a standstill. But with the new sprouting of leaves the dormancy is broken and the cambium lays down new xylem to add another growth ring to the over-expanding cylinder of wood.

It is obvious that winter-induced growth stoppages are trivial by comparison to a set, inborn stopping of growth. The difficulty comes, however, in separating the two, for often a fixed inherited stopping is somehow geared into seasonal changes. This is true of spores, sclerotia, pupae, and even whole insect societies such as the wintering bee colony.

Equilibrium at the point of maximum size

Size limits appear sporadically throughout the plant and animal kingdoms. In the first place, it is well known that in general unicellular organisms have a size limit. As Jennings (1920) and Adolph (1931) have stressed, the individuals may vary considerably, but the mean size of individuals in a clone is fixed and is not affected by artificial selection.

Many algae have size limits, though others do not. Such limits are poorly defined in *Spirogyra* and other simple filamentous forms, as well as in the larger thalloid forms such as *Porphyra*, *Ulva*, and the meristematic kelp such as *Laminaria*. In others, particularly the various colonial forms such as *Volvox* and its more primitive relatives, as well as *Pediastrum*, there is something that can be described as a partial size limit. In most of these cases the number of cells of the colony may be restricted within a given range, but if the cells can keep growing and enlarging, the colony size is not fixed. In *Volvox* the cells do reach a size limit while, in *Pediastrum* they keep expanding, although apparently within limits.

The fungi also show instances of continuous growth and instances of a size limit. This is particularly interesting because often these two properties are exhibited simultaneously by the same organism. For example, in *Aspergillus* or *Mucor* the mycelium seems to expand indefinitely provided the nutrient is available (as growth in race tubes clearly demonstrates), but the conidiophores or sporangiophores are quite consistently fixed in size. The difference is even more striking in the case of mushrooms and other compound fruiting bodies, for the whole structure will be quite consistent in size for a particular species (Plate 14). The original number of cells laid down in the primordium is roughly constant, and if cell expansion is complete the end result is a consistency of size (Bonner, Kane, and Levey, 1956). One of the interesting aspects of this example is that by the time the spores are formed, the whole mushroom (that is, the fruiting body itself, not the subterranean mycelium) is essentially dead and useless. If the spores are shed in one flush, as in the case of the delicate *Coprinus*, the fruiting body will quickly deliquesce and disappear; if the spores are shed over a period of time, the cap may provide food reserves to the spore-bearing gill, but once it is gone, the whole mushroom rots away. Therefore, this particular size limit is literally a dead end, and senescence follows very quickly upon the cessation of growth. The period of maximum size equilibrium here is rather short despite the fact that the organism has a strict size limit.

Among higher plants there is a sharp distinction between those that possess secondary growth and those that do not. In the latter case there is one spurt of growth, flowering, and fruit bearing; in fact it is basically the same as the situation in mushrooms. With secondary growth there is, in general, a disappearance of the size limit and what the botanist calls a "continued embryology."

If we turn to animals, undoubtedly some species of sponges have size limits, especially those that have a regular and symmetrical shape. On the other hand, some of the more disorganized conglomerations seem to expand without stopping. But this is only guessing on the basis of appearance, and no one to my knowledge has made a close study of size limits in sponges.

One of the best known cases of really extended size equilibrium and size limit is found among coelenterates. This is the maintenance of sea anemones in aquaria for periods up to ninety years (see Comfort, 1956). As was mentioned previously, in other cases the size limit is reached and recession follows, as Crowell (1953) showed for *Obelia* and *Campanularia*. Finally, there is the celebrated case of hydra, where Brien and Reniers-Decoen (1949) demonstrated in detail that it maintains its maximum size by continually growing at the hypostome end and continually regressing at the foot end; since the two processes are roughly equal, the size is constant. This is a perfect example of a steady state. It should be added parenthetically that in the colonial hydroids such as *Obelia* we are talking of the fixed size of the hydranth that is reached before regression, for the whole colony probably does not have any growth limit, the relation between the two being roughly the same as that of the sporangiophores to the mycelium in a bread mold such as *Mucor*.

There is a great wealth of examples of equilibria at periods of maximum size among arthropods. The prime reason lies in the exoskeleton, since the only way to escape this prison is by molting. In many arthropods, such as crustaceans and insects with incomplete metamorphosis (Plate 21), there is a periodic stopping with each step, so they might have been discussed in the previous section, but in many forms there is either a final molt or a size limit to the repeatedly molting animal. In insects with complete metamorphosis, the chitin armor of the adult firmly holds a constant size. There are some interesting exceptions such as the honey ants, where the workers that store the honey become gorged and swollen. However, in this case it is only the abdomen that expands by extension of the soft and elastic membranes between the hard plates; it is size increase by overeating and not by growth. At the other end of the spectrum there are those many cases where the adult does not feed at all, such as the male mosquito that lives entirely on the reserves built up during the larval period.

Many fish and reptiles have often been described as growing

continuously, but as numerous authors have pointed out, the growth rate falls off severely in later years. By contrast, mammals and birds have a specific size at which the growth rate falls to zero. The question of the length of this period of size maintenance is obviously completely involved with the problem of mortality because the equilibrium period is ended only by death. In turn the mortality rate is considerably influenced by the ecological conditions. At one extreme is violent competition or severe climatic catastrophe, when the great likelihood of death reduces the average life span. At the other extreme are the comforts and pampering of captivity, where senescence and decay alone may finally end the period. In the case of man, we have greatly extended the period of maturity by keeping ourselves in a curious form of self-captivity.

Conclusion

It has been my arbitrary object to subdivide life cycles into periods of size increase, size decrease, and size equilibrium. The basic argument is that size is of key significance to the organism as well as a convenient criterion for classifying and categorizing parts of the cycle.

Size equilibrium can occur at any stage of the life cycle. If it occurs at the point of minimum size it is usually a resistant stage, a spore or an egg, although resistant stages appear later also, such as seeds and insect diapause. Other pauses during size increase and at the point of maximum size are not resistant in nature.

Many organisms do not pause either at the beginning or the middle or the end of the life cycle, but always go directly on to the next stage. There are many instances where the period of size increase is repeatedly interrupted with pauses (e.g. the molting of insects). Also there are cases where many cycles will go without a pause, and then suddenly there will be one; a periodic pause after a series of cycles without pause (e.g. sexual stages after many asexual cycles in numerous algae, such as *Volvox*, Plate 9).

In considering all the size changes that have been discussed in this chapter, we must not become lost in the detailed examples.

The basic principle of size change and its application to parts of the life cycle is so simple that it can be overlooked. One glance at the Plates or a moment's reflection on the life cycle of any one organism is enough to see clearly the periods of size increase, size decrease, and size equilibrium.

We must now explain how these ideas are meaningful and useful. Thus far the main argument has been that size, like quantity in chemistry and physics, is a convenient measure for classification. Indeed this has proved to be the case, for it is easy to organize on the basis of size. But there are other more important issues which will be considered in the next two chapters. One is the question of the mechanics of the size changes; how do the organisms achieve these alterations? This means, in the terms used here, an investigation of the nature of the steps. The other is the matter of the relation of the steps and especially the size changes to evolution. It should ultimately be evident how the steps, size changes, life cycles, and evolution all fit together into one fabric.

4. The Steps

IN the beginning, steps were defined as chemical reactions which follow one another in a set sequence. This matter now needs more careful consideration, for to some extent this simplified definition brushes over some interesting and important problems.

First of all, in this definition no account is taken of whether the reaction occurs within a cell or between cells. The answer is, of course, both, for in multicellular organisms many reactions can occur among substances that are given off by the cells. Much of the discussion to follow here is particularly applicable to single cells, largely because multicellular organisms will be discussed in detail later.

Sequences of steps in cell systems are framed by three main conditions: (1) the nature of the reactants, (2) the quantity of the reactants, and (3) their distribution in space.

(1) The nature of the reactants is really the substance of modern biochemisty. It involves the activity of salts, water, oxygen, carbon dioxide, carbohydrates, fats, proteins, nucleic acids, and numerous other substances. Perhaps most important of all are the proteins that are enzymes, for they control the rate of the reactions that are the steps. The reactants then are the substances that are known (and no doubt some that are not known) to the biochemist.

(2) The amount of the reactants clearly also plays a significant part of any reaction or step, a fact evident from elementary chemistry.

(3) For many years the biochemist and the cell physiologist have appreciated the fact that the spatial distribution of the reactants is of prime importance in any reaction within a living cell. It is not enough to have the substances present; they must be brought into proper contact. The cell is not merely a fluid, but is also a highly ordered solid or semisolid. Now that the electron microscopist has joined efforts with the cell physiologist,

many of these spatial biochemical problems are being attacked and some solved. Besides these rigid means of spacing the reactants, it is also possible that they can be spaced in the fluid phase; for instance, diffusion forces are bound to be significant in the spacing of free-moving reactants.

Within the framework of these three conditions there is another important consideration. Many reactions or steps go on simultaneously within a living cell, and any reaction may in numerous ways affect neighboring ones. This has recently become a special concern of biochemists, for it has been shown that many reactions are in fact controlled by other reactions, and there is in any living biochemical system a maze of negative and positive feedbacks, direct and indirect inhibitions—in fact a wide variety of internal control systems. Steps, apparently for the most part, do not act in isolation but influence one another. It is perhaps more than influence, for their operations are deeply enmeshed one with another.

Some attempt has been made to discuss the immediate conditions which produce the steps, but let us look back one step further and ask how were the conditions achieved. This question could be asked in another way, for the conditions are themselves the result of steps, and therefore we are saying that since one set of steps is dependent upon the previous one, how did the previous one become properly set? This is an ancient problem, and the answer must be that all the steps have been continuous since the origin of life on earth. Despite its philosophical soundness, it unfortunately is not a particularly helpful reply, for it seems merely to push the answer into the remote past without really grappling with the problem.

The answer can be somewhat improved if it is put in terms of the life cycle. The steps are clearly not all different since the beginning of time, but keep repeating themselves in life cycles. This means that by fully understanding the sequence of steps within a life cycle one can at least understand part of the steps in terms of antecedent conditions. This is precisely what the biochemist and the developmental biologist attempt to do. However, there is a certain basic segment in this life cycle analysis, and that is the cell. The cell is at the initial and final end of the life cycle; the cycle never, for any organism, goes to a unit smaller or less complex than the cell. If the composition of this irreducible unit is to be understood in terms of steps, then indeed we must go back to the origin of life. Unfortunately it is a matter that occurred so long ago that it will be very difficult if not impossible to be sure we have divined the steps that led to the first cell.

The fact that the normal, living sequence of steps cannot be reduced below the level of the cell has many important consequences. It means that any one step or sequence, isolated from the cell, can only be part of the whole story. We do constantly abstract in this way, but when implications are drawn, the process we have ripped out of the cell, usually for experimental purposes, must be carefully put back in.

An attempt has been made here to steer a course between the detailed facts of biochemistry and cell physiology and the basic philosophical considerations. The first would soon lose us in detail and the second (and perhaps here the experimental scientist is at fault) is not always satisfying. The intermediate generalities have the advantage in that they organize and gather groups of facts in such a way that the facts have greater meaning, and one can see more deeply into their truth. The broader philosophical generalizations have a way of including so many facts that they lose their helpfulness; they become remote and inapplicable merely by their universality; they have lost contact with the problems that we want answered. However, this does not mean that in their own fashion they are not significant and important.

Such a philosophical point arises here, for in standard biology of today the sequence of steps would be called "causal." Biologists are dimly aware that there are philosophical problems connected with the notion of causality, but these are of no great interest to him. He uses the term in the simple sense that one set of conditions leads to another, and he would prefer to concentrate on identifying those conditions rather than worrying about the semantic problems of the word "cause." This, of course, has not been the case for the physicist, because with the advent of quantum mechanics he has been forced to reject the notion of causality. For instance, the time when a given radioactive nucleus disintegrates is intrinsically random; according to quantum theory no "cause" can be ascertained for the individual disintegration at some specific moment. The half-life of a quantity of radioactive material is then a statistical matter which can be ascertained by averaging the characteristics of many individual nuclei.

All biological steps are statistical in nature also; they are so in the same way that are physical and chemical steps, for indeed they *are* physical and chemical steps. But because the probability that they will go in a certain direction is so great, predictability becomes reliable and old notions of causality are as serviceable to the experimentalist as the more modern statistical ones. Here I have tried to evade the problem by using neutral, unloaded terms: steps, and sequences of steps.

Despite the problems connected with causality, our prime interest remains in understanding the conditions which produce the conditions, etc. The complete description of the sequence of steps remains the goal of the modern biologist. He appreciates fully that there are so many steps in any one life cycle that it will be doubtful whether he can identify them all, but at least he can find the major kinds of steps and the major categories of sequences.

Among biologists there has been a full awareness of these problems; for instance, Waddington (1940, 1957) has developed some special terms to cover the situation. He uses the word "epigenesis" for the sequence of steps and the word "canalization" to indicate the fact that the sequence is relatively rigid, like a ball rolling down a groove or valley. The only problem these terms present here is that they have been defined specifically for the period of development and not for the whole life cycle. Rather than alter their definition, I have preferred to use the more all-inclusive words "steps" and "sequence of steps," which can apply to all parts of the cycle.

Part of the discussion that follows will be on the identification of some of these categories of steps. As before, it will remain on a general level and not go into detailed examples and evidence. For instance, the process of gene action constitutes an important sequence of steps which has been successfully elucidated in recent years by the biochemical geneticist. Let us briefly examine this and try to put the important facts in their proper setting.

Gene action

The most abstract (and the most conventional) view of how genes act is the direct result of the extraordinarily recent advances in nucleic acid biochemistry. The DNA is coded within the nucleus, and with this code it makes more of itself during reproduction; to affect the cytoplasm it makes specific RNA which moves out of the nucleus, passes the information to cytoplasmic RNA, which in turn makes specific protein with the information given off by the nuclear DNA. The demonstration of this sequence is elegant and exciting. But long before these discoveries were made it was pointed out by numerous biologists that any such sequence was only meaningful if put back in the context of the cellular environment, and now recently Commoner (1964) has suggested in very specific ways how this can be done in the light of our new knowledge.

The point is a simple one, and in fact has already been made. The cell is the smallest irreducible unit in the life cycle, and therefore the minimum unit of inheritance is not just the genes, but the whole cell. We do not merely inherit coded information for blue eyes and blonde hair from our forefathers; in fact our foremost inheritance is cells. This obvious point has a number of interesting implications.

In the first place it means that genes cannot do anything by themselves, but only in the environment of the cell. After all, if the cell were not there, there would be nothing to direct; the genome would be like a general without his army.

Secondly it is well known that genes do not all fire off at once, but depending upon the cytoplasmic conditions, certain genes fire off and not others. This phenomenon has been elegantly demonstrated by the experiments of Beermann, Pavan, Clever, and others (review: Beermann, 1964), where chromosomes puff in specific tissues at specific times and on specific bands of the chromosome. This means that the local environment within the cell is informing the chromosome as to when and where to act. In fact, Commoner (1964) goes even further to suggest ways in which chemical information can be passed from the cell substances to the DNA of the chromosome. Regardless of the details and their final solution, we must include the internal environment along with the genes as contributing both substances and information to the gene action. There is no conceivable way for it to be otherwise.

This means that if we consider gene action during the course of the life cycle of a single cell, such as a bacterium, the cytoplasm is of such a nature that it elicits genic activity which changes the cytoplasm which again alters the genes. The steps then involve both chromosomal DNA and the surrounding protoplasm. Furthermore, as stressed before, there can be numerous such series of steps going on simultaneously in different parts of the cells; it is not just one series of steps.

It has also been pointed out that steps affect one another, and there is ample evidence to illustrate the point. First of all, on the gene level, one gene may suppress or enhance the effect of another. Secondly, no one gene is likely to have one effect, but many, as has been clearly demonstrated in the many studies of pleiotropy. Not only that, but one gene may modify the effect of another by suppressing some of its pleiotropic manifestations and allowing others to express themselves. All these permutations are presumed to be important elements in adaptation and natural selection, as Mayr (1963) has recently emphasized.

Moreover, the cell is the unit of inheritance in another sense also. As has been known for a long time (and discussed recently in detail by Sonneborn, 1963), there are many cytoplasmic structures which are simply passed on directly from generation to generation, and some of these, at least, may not be capable of synthesis *de novo*. For instance, Ephrussi (review: 1953) demonstrated that certain mitochondrial deficiencies cannot be regained unless they are reintroduced by cytoplasmic exchange with a normal individual. Some of these cytoplasmic structures which are passed on are known to have some genic control as well, but even if the gene favoring the structure is present, the structure cannot be created if it is absent in the cytoplasm; the gene merely favors the reproduction of existing particles. Therefore in each generation the organism inherits directly many of its cytoplasmic components. If they are missing then they cannot be replaced, and if they are essential, death is the inevitable result. The cell is the package that carries all the structures and particles necessary for the next generation, including the genes on the chromosomes.

The end result is a vast network of control systems that involve both the genic material and the cytoplasmic substance; it is a thick, interconnected meshwork. Parts of it have, by brilliant experimental researches, been dissected out, but such parts can operate only in a whole unit, and the lowest common denominator is the cell. The unit of inheritance is certainly not the gene; the gene is only part of the cell, which is the true unit of inheritance. This unorthodox statement, however, is correct only if we mean by unit of inheritance the minimum unit that can carry on the heritable information from one generation to the next.

It should finally be noted that the minimum unit is the point of minimum size in the life cycle; it is therefore minimum and irreducible in all senses, and is the link between one life cycle and the next.

Internal versus external stimuli

We have been discussing at some length the fact that cells have internal stimuli and responses in the form of an internally controlled series of steps. It is also clear that organisms respond to the environment in various ways, and now the question is what is the relation between the internal stimuli and the external or environmental ones.

One class of external effects is too coarse to be of any significance in this discussion. For instance, the temperature during which size increase occurs must be at a physiological level. If it is too cold or too hot the organism will either remain in suspended animation or die. Temperature, in this case, can hardly be called a stimulus, but rather an environmental condition requisite for the development of the individual.

Another type of external stimulus of greater interest is that in which the environment gives a rhythmical cue. In recent years there has been much work on biological rhythms, and certain activities of animals and plants are geared to diurnal, lunar and annual cycles. In these cases it has been shown that changes in light intensities, which are invariably associated with these environmental rhythms, usually act as stimuli upon the organism so that the cycle of the organism will correspond exactly with that of the environment. In many cases of diurnal rhythms, for instance, dawn is shown to be the signal to keep the external and internal cycles synchronous. This may govern, for different organisms, periods of rest and activity which may be associated with feeding, reproduction, and other functions. In all these instances the organism is sensitive to a change in the environment and acts accordingly.

It is interesting that these rhythms are mostly endogenous, and continue even if the organisms are brought into artificially constant conditions. Often the duration of the cycle is not quite equal to the solar or lunar rhythm, but it is close and is apparently constantly re-set by the external rhythm. This internal endogenous rhythm is not just present in higher animals with complex nervous systems, but also in plants (see Bünning, 1964) and in unicellular organisms such as *Euglena* (Bruce and Pittendrigh, 1956). It is true that there are some cases where the entire rhythm is imposed by the environment, but it is remarkably common to find that they are internally controlled as well.

It is obvious that all these rhythms are of considerable adaptive significance for the organism, but it is nevertheless puzzling that in so many instances there should be two separate controls of the rhythmic activities of the organism. Nature usually does things in the most efficient way possible, and it seems strange that in so many of these cases there is both an internal and an external stimulus where the external one would have seemed sufficient. One possible assumption is that the genetically controlled response to external clues came first during the course of evolution, and that its internal genetic fixation followed as a further adaptive safeguard. Undoubtedly all these effects are polygenic and therefore it is possible to have a complex but reliable situation where the rhythms can occur independently without the environmental stimulus, yet with the addition of an external stimulus they can be set or timed so that the organism is not only in tune with the outside world but can manage when the outside fails to send (or the organism fails to receive) its customary signals.

A very similar situation arises in the so-called "Baldwin effect" (review: Simpson, 1953) where a tissue will respond to environmental conditions, such as the callosities of an ostrich or the soles of our feet which grow thicker with abrasion and use, yet these same tissues already appear thick before birth, before any abrasion was possible. Here also a particular end result is achieved both by external and internal stimuli. Waddington (review: 1957) has done some very interesting experiments upon *Drosophila*, where he was able to select for a response to a temperature shock (cross-veinless wings) to the point where the response occurred without the shock. There are apparently many genes involved in the determination of the character and others that affect their degree of expression, and by selection it is possible to produce the direct result of cross-veinless without the external stimulus. The sensitivity of the response can be increased to the point where the genes for cross-veinless act independently of the genetic system involved in the temperature-sensitive trigger. For our purpose the important point is that stimuli, in some cases for the same response, can emanate from either the external or internal environment of an organism. Furthermore, and this is the particularly significant part of Waddington's contribution, it is possible to separate by selection the genetic mechanism that controls the internal effect from the one that permits the organism to make an appropriate response to the environmental clues.

One final point about internal and external stimuli: with an increase in morphological complexity of the organism there is an increased dependence upon internal stimuli. This can best be illustrated by showing two extremes. Assuming an optimal diet, the rate of size increase of a fungus is entirely dependent upon the external temperature and humidity. It is possible to grow a gram of mycelium in a few days or a few months, depending upon the conditions. Each species will also have its intrinsic rate, but this is in turn affected by the environment.

At the opposite end of the scale is a placental mammal, where the embryo is kept in the constant temperature and food and oxygen supply of the womb. Here every effort seems to be made to shield the embryo from the raw outside world. Nothing could be more insulated from the outside than total enclosure in a homeothermic animal. All the stimuli are from within and are independent of external environmental changes. Nor does this abruptly change at birth, for with parental care the later period of size increase is still to a large extent protected.

If we turn to societies, possibly because the individual members of the society are not attached but fit loosely together, the case is not quite so striking as with mammals. Nevertheless there are some impressive examples that confirm the basic generalization. The best instances are among the social insects; I shall cite two. Termites build elaborate tunnels through wood, but even when their avenues pass over inedible materials such as concrete or metal they build covered tunnels of sawdust. At first it might be assumed that this is due to abhorrence of light or eagerness to stay hidden from predators, but Lüscher (1961b) has shown that the prime function is to create an internal environment of constant humidity. In a particularly interesting series of studies Lindaur (1961) has shown that honey bees are adept at maintaining constant hive temperatures. In cold weather air circulation is restricted, while in excessively hot weather water is regurgitated by some workers while others fan vigorously, permitting a considerable cooling effect by evaporation. By these methods the bees manage to keep their hive temperature within a range of 34.5 to 35.5°C despite extreme external temperatures up to 70°C. In both these cases there is an effort to produce a constancy of the *milieu interieur*, to use Claude Bernard's phrase, and as a result there is automatically a reduction in the dependence upon external cues for the steps, with an increase of stress upon the internal ones. As with mammals we may assume that complex processes involving size-increase under these circumstances will undoubtedly be more precisely carried out than if they were exposed to all the vicissitudes of the outside environment.

Conclusion

The cell is the unit of inheritance (and not just the genome), for the cell is the irreducible unit of inheritance in the life cycle of all organisms; it is also the point of minimum size.

Within this minimum package there are massive intermeshing sets of steps which involve the genes and all the other parts of the cell. Stimuli and responses, actions and reactions of a controlled nature weave back and forth among these cell components so that gene action can occur.

Stimuli or triggers for any one reaction may stem from the environment or may be internal. Each can be adaptive in its own way, but often a particular reaction will have both external and internal triggering potentialities. Perhaps this is just a way of ensuring harmony between internal and external events. Furthermore the generalization may be made that single cells are usually far more dependent upon environmental triggers than are multicellular individuals. To put this in another way, size increase brings with it relative independence of the environment, although larger organisms still depend upon the environment for cyclic cues of climatic changes.

Thus far we have concentrated primarily on the case of the single nucleus with its surrounding cytoplasm. It is now time to turn to one of the most intriguing questions in biology: How can these same kinds of steps take place in multicellular organisms so that controlled differentiation and division of labor occur? The cell cycles are subordinated within the life cycle of the multicellular organism, yet the cells in different parts take on (with rigid consistency from generation to generation) different characters in a beautifully organized fashion.

THE STEPS DURING SIZE INCREASE

Gene action in populations of associated identical genomes

Let us for the moment assume that the genomes are identical in all the cells of the multicellular organism. If they are identical, how can differences arise in different parts? If they are the same, would not one expect the parts to be identical also? There are some analogies in the non-living world that illustrate the point. A crystal may be made up of identical molecules, but because of the way the molecules are stacked, different faces may have different properties; its faces are in fact differentiated.

At the other end of the scale, even a population of individuals will have at least similar genomes, although they would not be identical. (This statement refers to all the genomes of each cell of each multicellular organism within the whole population.) In certain insect societies (e.g. Hymenoptera) and colonies of multicellular organisms, the genomes are identical (neglecting possible somatic mutation), for they are all derived from a single zygote nucleus. In each of these groupings the distinguishing character is the degree of closeness of the nuclear association (as well as magnitude and the number of nuclei). But the problem is the same for all; it is the old familiar problem of how it is possible for a mass of genetically identical nuclei to produce differentiation, either within an individual or in a society. How can one have a collection of identical things, and out of it get all the beautifully organized groupings we call organisms or societies?

The intimacy of the association of the similar nuclei will have profound effects on the degree and precision of communication between cells. It is self-evident that a population of unicellular organisms cannot perform the kinds of complicated differentiations which are characteristic of even the simplest multicellular forms. We assume that this is largely because in the latter case messenger substances or chemical signals can be readily passed from cell to cell, while if there are larger distances between cells (and perhaps changing distances if the cells are motile) the controlled passing of information is far more haphazard. Furthermore cells in contact can pass large molecules such as proteins and possibly nucleic acids that can be relatively high in information content, and this is impossible if the cells are even a small distance apart.

The principal kind of signal, at least on the cellular level, is the passing of chemical substances. This may be a haphazard process, following the random laws of diffusion, or it may be directional. In the latter case, by some kind of structural polar orientation, substances pass more readily in one direction than in others. The classic example is that of the polar transport of auxin in vascular plants, but there are undoubtedly many others.

The whole question of polarity and directional signals is poorly understood. In the next section we shall consider some of the special properties of polar arrangements, but it is clear that directional alignment in itself can be used as a means of passing signals between nuclei, the polar transport of substances being a special case.

Even though our ignorance is great concerning both physical and chemical signals (for despite their importance they are difficult to pin down by experiment), we still can appreciate their great significance in communication between nuclei. We have a better understanding of the problem of communication in multi-cellular animals. There special organs or structures exist to emit and record physical signals such as sound vibrations, light waves, electrical fields, and chemical signals such as odors and tastes. The result is that separate multicellular individuals, with their nervous systems and end organs, can become integrated in all sorts of interesting ways even though they may not be in direct contact with one another. (For a discussion of signalling in animals see Tinbergen, 1951, 1953; Marler, 1959; Haldane, 1954; Wynne-Edwards, 1962.) Among higher plants most of these intricate signal systems are absent, though many plants are known to give off substances that either attract, repel, inhibit, or stimulate the growth of other individuals at a distance (e.g. James Bonner, 1949).

At first it may seem far-fetched to consider these as special cases of communication between nuclei. If a flock of birds comes together in a tight bunch, can we say that this particular bit of social organization is due to the signals passed between the many millions of nuclei flying through the air! In a very simple though indirect sense the answer is definitely yes. The genes are responsible for the instinctive wing movements which are the visual cues recorded by the eyes and to which the birds respond instinctively by bunching. The structure of each bird is gene-directed in its development, and so undoubtedly are the instinctive actions, as may be inferred from numerous breeding experiments on behavior patterns in animals. It is true that there is the possibility that some learning is also involved in flocking, but even the ability to learn probably has a genetic basis. In any event the flocking is in one way or another affected by the genome, and the result is that the birds come close to members of their own species under certain seasonal or daily circumstances (i.e. external stimuli). Therefore we see that genes, by governing instinctive behavior of organisms that have good signal manufacturing and receiving instruments, can control all of the social organization. This much is clear and straightforward; what is more difficult to grasp is how this is possible if all the genes of all the nuclei are the same or at least very similar. Again, how can one get form and differentiated structure from a mass of identical units?

We must first consider whether the nuclei of an organism or in a population are in fact identical. In embryology this is an old problem and one which continues to be an area of active research. Perhaps it remains such a difficult problem because in some cases the nuclei remain identical and in others they do not. Let us examine a few cases to see how they compare in their role in multicellular or social differentiation.

The clone of a unicellular organism, assuming that no mutation has occurred, provides the basic example of the case where all the nuclei are the same. It is also true, of course, that there is usually little or no differentiation in this case. It is possible, in a bacterial colony, that those cells on the inside of the mass will have a different chemical environment from those on the edge and as a result may differ in appearance and shape, but since any one of these cells can start an identical new clone, the nuclei are considered equal.

Among complex organisms, the best known case of genetically equivalent nuclei is that of higher plants. Among flowering plants there is considerable evidence that a single cell from a root callus grown in culture will ultimately produce a whole new plant (Steward et al., 1961; Torrey, 1963). Among multicellular animals the situation is not quite so striking, although it is known that a number of invertebrates such as hydroids, planarians, ascidians, and various others are capable of regeneration from very small groups of cells isolated from different parts of the body. It is still possible, in these examples, that each cell mass contains a number of genetically diverse nuclei.

The most celebrated study of this sort is the fine work of Briggs and King (review: 1955) on the substitution in enucleate eggs of nuclei from later embryonic cells. But despite these elegant experiments and those of Fischberg, Gurdon, Moore, and a number of others, the case is still in doubt (see Gurdon, 1963). In some instances the gastrula nuclei show fixed abnormalities that are handed down through a series of transfers, and in others, such as the transplantation of kidney sarcoma cells, whole, perfect embryos result. In the first instance it is not clear if the altered nuclei (which frequently have an altered karyotype) arose by damage in the transfer or whether the kidney sarcoma cells have genetic properties that normal cells lack.

There are, however, cases where there is a clear nuclear alteration. The classic one is that of *Ascaris* studied by Boveri (1887 *et seq*. See E. B. Wilson, 1924) where the germ line nuclei retain two large chromosomes, while those of the soma fragment into many small pieces. In fact it could be said that this type of example, where there is a clear differentiation between germ and soma, is the only type of nuclear alteration during size increase that is established without any shadow of doubt. But the nuclear differentiation that can be demonstrated is of a relatively uninteresting sort: some nuclei can produce entire new individuals (germ), while others have lost this ability (soma). The fact that in the latter case the ability is lost makes it impossible to determine if the nuclear change is trivial or of fundamental significance.

There are a number of instances where animal tissue cultures have been initiated by a single cell, and if the conditions of the medium are favorable there will be only one cell type (such as muscle cells, Konigsberg, 1961). However, these examples do not tell us very much about the composition of the nucleus other than its ability to foster a particular cell type. Rather their importance lies in showing what kinds of conditions favor cell differentiation in a certain direction, and along with this all the studies on induction in tissue culture have added greatly to our appreciation of the specialized localized conditions within an organism which promote or trigger a particular type of differentiation. The nucleus of a nerve cell is not necessarily different from that of a muscle cell.

Among social insects there is an apparently similar instance of nuclear differentiation, as Wheeler (1911) pointed out some years ago. Only the reproductives are capable of having descendants; all the workers are sterile. Again, however, this is a trivial difference, and in fact among termites if the king and queen die or are removed, secondary reproductives will arise in succeeding molts from sterile workers, showing that they have not lost the potentiality to produce new families, but this ability has been continuously suppressed by ecto- or social hormones given off by the king and queen (Lüscher, 1961a). Therefore in this case the egg and sperm nuclei of all the individuals in a society are again clearly equivalent; the differences are produced by local environmental conditions.

In all other types of societies and multicellular populations there are genetic differences between individuals, but these play no role in the structure of the population or social group. The variation is part of the mechanism of natural selection and has nothing *per se* to do with the stratification and composition of the social group. There is, however, one great exception to this sweeping statement. Sex differences are genetically determined, and most animal societies are strongly influenced by sex differences. The sexes may have different structures and functions within the group other than those directly connected with the production of eggs and sperm, and these play an important role in the differentiation of the social group.

Despite this example of sexual differentiation and the possible instances of nuclear alteration during embryonic development, the significance of genome differentiation in the life cycle is severely limited. The reasons for this are twofold: In the first place, it is clear that in those cases where there is no alteration of the genome there can be complex differentiation, as in higher plants. In the second place, even in those cases where genetic differences are certain, as in sex determination, this is still only one small step, one small contribution towards producing an organized differentiated structure. The real problems have no such simple answers, and in fact it is best to clear the slate and momentarily forget even the suggestion that differentiation of the genome might occasionally play a part in the differentiation of organisms. Then we can come to grips, as we shall now, with really fundamental matters.

Among the larger living systems, particularly during their period of size increase or at least during the maintenance of their maximum size, there are four types of organization that have particular significance, each of which will be examined separately. The first (*i*) is the simplest case, in which there is no obvious differentiation and all the individuals are apparently the same. The second (*ii*) is the case where there is a rigid partitioning; the cells or the organisms cannot move with respect to one another and yet they somehow manage to become highly differentiated. The third case (*iii*) is where the cells or organisms can move and can take up various positions so that the whole group has an elaborate and organized structure. The fourth and last case (*iv*) is where all the cells or parts together go through a series of changes in time, the end result of which is a differentiation if one considers all the functions that have been performed during the whole life cycle. Case *iv* is unlike conventional differentiation; here the differences or divisions of labor appear serially, while in conventional differentiation they appear simultaneously. In all the cases where there is clear differentiation (*ii*, *iii*, and *iv*) we particularly want to know how the order is produced in this population of similar if not identical genomes, and furthermore to examine, in all these cases, the effect of size on the nature of the sequence of steps.

No differentiation among the units

If there is no differentiation, all the units are the same. However, it is undoubtedly incorrect to refer to any group of cells or individual multicellular organisms as being the same, for this connotes an identity of structure far beyond what could be demonstrated or beyond what probably exists. Rather what is meant here is that the units are roughly similar, and even if upon careful inspection they show slight differences, these variations are random and do not constitute any kind of organized differentiation.

At first glance it would appear that if all the units are the same, there could be no internal structure in their grouping. This, however, is far from the case; as was already mentioned, the best example does not lie in the province of cells or organisms but at the lower level of crystals. In this case the units, the molecules (or unit cells), are truly identical, but as a result of their surface forces they become aligned in a precise and orderly way. Thus the crystal has faces that orient with respect to one another at definite angles, giving a perfect and predictable form to the aggregate of molecules. Furthermore there is differentiation as well, for some of the faces of certain crystals will have quite different properties from others because, with the orientation of the molecules, the atomic groups on one side of the ranks of molecules will be different from those on the other. All these collective properties can be entirely accounted for on the basis of the attraction forces which bind the molecules together.

If we turn to cells it is first of all clear that, like molecules, they can have a front and a hind end. If they are polar, and if they are aligned so that the polar axis of each is pointing in the same direction, then one would expect different properties on one side than another.

While this is formally true, we immediately have difficulty when we try to provide examples. In the aggregating amoebae of the cellular slime molds, Shaffer (1962) has shown that the cells hook on to one another in an end-to-end fashion, and he has suggested that they retain this position in the cell mass. Further he has argued that the fact that the front and hind ends of the cell mass have different properties might be, at least in part, ascribed to this aligning of individual polarities. This cannot be the whole basis of the regional differences, as we shall see presently, but it may indeed play a significant part. Unfortunately it has not yet been possible to demonstrate precisely what part. It is undoubtedly true that the cells are aligned and that the polarity of the whole multicellular organism is the direct result of this alignment. The alignment arose partly by the acrasin chemotaxis and partly by the specific contact adhesion zones described by Shaffer. The resultant polarity of the cell mass clearly influences, if not causes, many aspects of the subsequent development. Exactly how this occurs will be a fruitful problem for future experimental research.

Among aggregative organisms, another fine example of the significance of cell alignment may be found in the myxobacteria. Here the equivalent rod-shaped cells stream in one direction and ultimately form a polar fruiting body which rises into the air. The organization of this structure is obviously dependent upon the cell alignment. And as a result of this directional movement of the cells, the final fruiting body of any stalked form, such as *Chondromyces*, has a top and a bottom to it; it has a communal polar organization (Plate 5). The advantage of this particular example in the present context is the fact that the cells all differentiate into spores or cysts; that is, they are all equivalent and do the same thing.

If we look among truly multicellular plants and animals for

examples of individuals made up of identical cells, we find very few. In the vast majority there is elaborate regional differentiation, and the only exceptions are among some of the lower plants, especially the algae where there are straight filaments (e.g. *Spirogyra*), rectangles (e.g. *Merismopedia*), cubes (e.g. *Eucapsis*), and spheres (e.g. *Pandorina*) (see Fig. 4). In each of these, while the cells are apparently equivalent, the polarity of each cell of the group is rigid; and even more important, the relation of the cell polarities within the group is rigid so that the whole multicellular organism or colony has a fixed pattern. This communal polarity is apparently caused by the rigidity of the bonds connecting the cells and the direction of growth. In some cases, such as a straight filament of cells, each cell can grow only in one direction. However, in rectangles such as *Merismopedia* all the cells alternate their directions of growth at successive right angles in a plane. We must assume that in this case there are systems of communication between the cells which synchronize the timing of this growth. The end result, irrespective of the extent of the communication from cell to cell, is an organized pattern resulting from the adherence of identical cells. At least in this respect these simple algae parallel some of the properties of crystals.

The next place in our size scale where we find groups of similar units is among multicellular plant and animal populations. It may be said as a general rule that in those cases where the individuals are similar the amount of group geometric organization or pattern is rudimentary or nonexistent. A field of grass or a forest of pine trees does not seem to show much organization or pattern. However, among animals there are some interesting although ephemeral cases of order. A flock of birds or a school of fish may have shape to it, although the shape may be to varying degrees transitory (like flow birefringence). But the shape is entirely the result of the fact that the birds or the fish are all aligned and going in the same direction. As is well known, they achieve this orientation through visual signals, and the result is a wonderful organized parade in which the group as a whole will have a front and a hind end.

But the important point is that again the organization comes from the aligning of the individual polarities, and without this there are varying degrees of disorder. Therefore similar units from crystals to bird-flocks can achieve (without differentiation) some over-all organization by marshalling the individual polarities into the polarity of the whole grouping. Polarity is, then, the most primitive method of coordinating a group of small units into one larger unit. But combined with polarity there may be other methods resulting in even more elaborate organization and differentiation of parts.

Differentiation among rigidly spaced units

We now come to those cases where the polarity and structure of the group of cells or individuals are rigid (as for instance with crystals or *Spirogyra*). The units cannot move about with respect to one another, yet some of the units become different from others. Plants as a group provide perfect examples of this kind of organization. Having hard cell walls the cells cannot slip about with respect to one another, and yet if we look at a flowering plant, the cells are differentiated into roots, shoots, leaves, etc. Since all these differentiations arise during the period of size increase, this is largely a problem of development. We must now ask how the identical genes achieve this remarkable bit of epigenesis.

There are at least four different ways this can occur, and each will be considered briefly with a few selected examples, but first I want to emphasize that any one organism can use more than one of these systems during the course of its development.

The first method of differentiation among rigidly spaced units is in some ways the most obvious. In this case the whole pattern is simply passed down directly from the previous cycle. The classic case is that of ciliates, where all the elaborate structure of the cortex, the basal granules, and all their delicate affiliated structures are never destroyed but are kept intact continuously from cycle to cycle.

Even when sexual fusion occurs there is no loss of the cortex or a reappearance of a new one, for the two conjugants merely exchange nuclei and then separate again. After conjugation the old macronucleus disintegrates and a new one arises from a recombined micronucleus. This in turn can give new orders, new messages to the cortex. Any alteration achieved in this way is then passed on directly through the succeeding cell cycles. It is also possible in this way to pass on cortical abnormalities that have no nuclear origin; many authors have pointed out that the passage of cortical information from mother to daughter in the ciliates is in part a cytoplasmic inheritance as well.

It should be added that some new structures do arise after cell division. New contractile vacuoles, oral structures, and various other parts come into being in a bizarre and fascinating variety of ways before and during cell division (see Fauré-Fremiet, 1948). Here lie some of the really interesting problems of ciliate development, for the initiation and the organization of these organelles has some most intriguing aspects. Tartar (review: 1961, 1962) for instance has shown in *Stentor*, that zones where there are broad stripes lying next to narrow stripes, that is zones of "stripe contrast," induce the formation of the oral groove, indicating that not everything on the ciliate surface is pre-formed and inherited, but that there are also some inductive steps involved in the formation of new structures (Fig. 9). How-

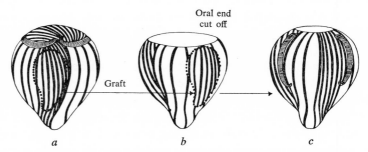

FIGURE 9. Induction by stripe contrast in stentor (*S. coeruleus*). A section of fine striping has been grafted from animal *a* to the flank of animal *b*, so that now there are two regions where broad stripes lie next to narrow stripes, and both of these zones of stripe contrast induce oral structures as shown in *c*. (After Tartar)

ever these new structures are laid down rigidly in the pattern inherited from the previous cell cycle.

The second method of creating differentiation of parts in a rigid system is beautifully illustrated in the development of many mosaic invertebrate eggs. Here, as shown in Conklin's classic work (1905) on the ascidian *Styela*, there are evident movements of the cytoplasm following fertilization and preceding cleavage. In brief, to summarize many beautiful observations and experiments, the substances in the cytoplasm are relatively homogeneous or random at first, and subsequently they become localized in specific regions, perhaps because, as Costello (1948) has suggested, diffusion gradients are set up by the puncturing of the egg by the sperm at one spot. In any event, specific "organ forming substances" (to use an old phrase) are localized in specific regions, and with cleavage these regions become cells and tissue of specific structures. Removal of a region of cytoplasm containing the future mesoderm will result in an abnormal embryo lacking muscle, or the nature of the substance can be demonstrated by centrifuging the cell and redistributing the substances before cleavage (Conklin, 1931). It should be added, as the work of Rose (1939), Reverberi (review: 1961), and others have shown, that in later stages there is also some induction, and the whole differentiation of the ascidian is not established in this way. Furthermore the egg or the adult show remarkable powers of regulation and are not in any sense mosaic (Dalcq, 1941). However, for the principal period of size increase, the ascidian has avoided the problem of rigid cells by moving the cytoplasm in a definite pattern and then chopping this pattern up, by cleavage, into cells.

The third method of making the units of a rigid system different depends on the relation of an organism to its environment as C. M. Child (1941) showed so clearly. For example, if a mass of cells is sitting on the bottom of a dish (or the ocean floor) the lower cells would be relatively low in oxygen availability and high in carbon dioxide accumulation, at least when compared to the uppermost cells. This difference could lead to regional differences in metabolic activity which in turn could lead to differences in chemical structure that could be called proper differentiations. Child's own work and that of Barth (1940) and many others on the regeneration of the hydroid *Tubularia* have supported this general conception in considerable detail, and there are many other good examples from invertebrates (see Child, 1941).

The fourth and final method that will be discussed here is perhaps the most common and the most advanced. Here we are concerned with those cases in which there is a group of identical units which by some internal means of communication manage to divide the labor. The first step may have been induced by environmental differences (our third method), or there may be something inherent in the polarity of the system which was inherited (our first method), but from this initial directional frame it is possible to produce all sorts of internal changes.

Herein lie many of the modern problems of experimental embryology. If the cells are rigidly fixed in relation to one another, but if there are gradients or avenues of polar movement (as in higher plants), it is possible for the parts of the cell-mass to communicate with each other. Stimuli and responses are no longer general for the whole mass, but regional (and therefore relatively specific), and each region may further break up into subregions. The whole process of induction is a key step of the stimulus-response systems of this sort; one region, by virtue of a gradient or of the polar structure or for some other reason, will produce a stimulus, and the tissue in its neighborhood will respond in a particular fashion. The responding tissue may be special and specific because of a previous condition: a gradient, a polar accumulation of some substance, which amounts to a previous induction. By the succession of each of these dependent and interlocking steps, the organism becomes more and more differentiated. With increase in size, the number of these steps is increased and the result is a greater and more elaborate division of labor.

All four of these methods may work together in one organism. Since the cells cannot move, the movement must take place before the cells are properly formed (oöplasmic segregation), and/or the predisposition of parts must be directly inherited, and/or the environment must cause regional differences which produce gradients, and/or the polar movement of substances must produce organized bits of cell-to-cell communication which result in differentiation. Each of these methods is composed of sequential steps that can occur without any mass movement or redistribution of cells.

Differentiation among mixing and mobile units

Here we are dealing with a wider variety of possible units for

cell populations: multicellular organisms which have morphogenetic movements as well as populations and societies of multicellular forms—all may be included. First we shall examine
how the differences arise and then how they arrange themselves
in some sort of order.

We are concerned here with variations of a special sort. For
instance, if a certain character is genetically fixed, and if there
are no mutant variations in the population, then all the individuals should be absolutely identical with respect to this character.
The fact that they do not achieve this ideal has been pointed
out by a number of people; Waddington (1957) calls this kind
of variation "developmental noise." As he indicates, it has been
demonstrated clearly in the variation in bristle number in
Drosophila, and he considers it primarily the result of "im-
precision in developmental processes which involve large numbers of cells." This kind of variation is certainly not confined to
large organisms for if the character under study is size in some
protozoan, then, as Jennings (1920) and Adolph (1931) have
shown, there may be considerable variation in the size of the
individuals in a clone. When extensive selection experiments
were made (by Jennings and especially by Ackert, 1916) in
which the smallest and the largest cells were removed and separately recloned repeatedly, after a series of such isolations the
span of sizes given by the clonal offspring was the same as that
of the original parent (Fig. 10).

As I have pointed out elsewhere (Bonner, 1959a) what apparently is inherited is the ability to vary within certain limits.
The variation is therefore not genetic variation in the conventional sense, for the range of variation is genetically determined
but the size of any individual within that range is not. In order
to have a convenient label we shall call this kind of variation
"range variation." Besides size, there are of course many other
characters which are capable of quantitative differences and also
show range variation.

What service does this range variation perform in development? At first there would seem little advantage to *Paramecium*,
for instance, to vary in this way. But before making any final
judgment it would be well to consider some other cases where
a possible use of the variation is more obvious. In the cellular

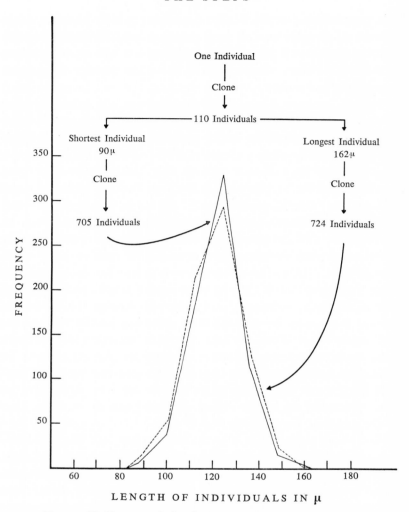

FIGURE 10. Range variation in *Paramecium aurelia* (from the data from Ackert, 1916). One individual began a clone of 110 individuals, and from this group the largest and smallest were selected. These were in turn cloned and their progeny plotted. Note that the parental size difference was not inherited and the size range is the same for both.

slime molds, the cells in a clone vary in size as well as in stalk- or spore-forming tendencies. During and just after aggregation of the amoebae, these genetically identical but somatically variable populations of cells are first aligned into a polar mass and, once given this order of direction, they proceed to sort themselves out so that the cells with the most pronounced stalk tendencies are at the front and those with spore tendencies fall behind (Bonner, 1959b, Takeuchi, 1963). Furthermore we must presume that these vary continuously and form a graded series in the polar cell mass, yet it is clear that ultimately a sharp division is drawn and there is a definite and proportionate relation between the number of stalk cells and the number of spores (Bonner and Slifkin, 1949; Bonner, 1957). Even if the cell mass is cut into sections, each section produces a proportionate fruiting body, as Raper (1941) originally showed, and from this we conclude that even a section of the gradient is capable of producing proportionate development of stalk and spore cells.

This example is emphasized because here the control of the steps is apparently achieved at least in part by range variation. The fact that the cells differ permits them (with the help of the alignment mechanism) to sort themselves out in a polar order. We do not fully understand how, but once they are in an aligned graded series they can block out proportionate development and put the division line between spore and stalk in the correct place. This precise control is achieved by a group of cells coming together by aggregation; but the significant quality is their range variation. It is this variation that provides the currency with which to make an organized pattern.

My original assumption was that these views were restricted in their application to slime molds and other organisms with morphogenetic movements, but it was of great interest to learn that Wynne-Edwards (1962) had come to closely similar conclusions in his considerations of mechanisms which control animal populations. His thesis, developed in detail in his important book, is that the size of animal populations is not merely determined by external causes but also by significant internal mechanisms that keep the number within viable limits and are therefore of great adaptive significance. If the population reaches an especially low or high number of individuals, this may be dangerous to the point of extinction. He suggests a number of different ways in which the population is controlled, especially ways which permit some degree of flexibility and reserve so that sudden increases may be quickly produced to make up for sudden losses, and vice versa. In particular he points out that the celebrated peck-order of Schjelderup-Ebbe and all other forms of social competition serve this function. In times of plenty all the individuals of a group, both the dominant and the dominated, will have ample food and apparently more opportunity to reproduce. In times of need only the dominant ones have the privileges. The social hierarchy has made it possible to keep a stable core of individuals in prime condition, even under unfavorable circumstances. Often the competition is for sites or territories, and success in obtaining food or reproductive success are secondarily associated with the territories.

To make a comparison with slime molds, it should first be mentioned that in all breeding experiments on dominance in birds and mammals there is no evidence of any kind of genetic determination of social position. It is reasonable to assume that social position is a range variation, at least in part. This most certainly is the case in the social insects where there is much evidence that in caste determination the individuals are genetically similar, but the phenotypic differences are the result of external chemical influences in the form of ectohormones and special food. Admittedly this is a most elaborate form of range variation, but nevertheless the order of the insect colony is to a large extent dependent upon the range of castes. Range variation applies to both multicellular animals and animal societies.

In both cases the variation provides a structure in the group. The individuals do not just vary, but by virtue of their variation they have certain positional relations to one another. Moreover, in both cases this graded series of variants has a sharp cut-off. In the slime molds it is the line between spore and stalk cells; in higher animal societies, according to Wynne-Edwards' scheme, it is the line between the haves and the have-nots.

To return briefly to the population of *Paramecia* (or any other microorganism), one might again ask what is the function of the range variation here, or can it be dismissed as unimportant. It would certainly seem that no hasty answer is indicated, and in particular it would be interesting to study these populations with the ideas of Wynne-Edwards in mind. In some species of ciliates starvation causes the group to divide (by phenotypic variation) into two distinct classes, namely cannibals (or predators) and prey (Giese, 1938). Perhaps this is an extreme example of the use by microorganisms of range variation in order to maintain sufficient population to tide them over hard times. If this is so, then how does this apply to *Paramecium*? Is there simply a rough division between the haves and the have-nots as among the higher vertebrates?

Before leaving the subject of range variation a key point should be stressed. It revolves around the question of why the individual variants are not genetically determined rather than simply the range; what is the advantage in having this unconventional and relatively haphazard form of inheritance where genetically identical individuals can differ phenotypically over a considerable spectrum?

The answer is clear: if this were not the case it would be immediately impossible to use the variation for the kind of organizational hierarchy that we are suggesting here. In the case of peck-order, if the degree of dominance were inherited, there would be a strong selection pressure favoring dominant individuals. One could easily imagine that after a period of such selection the individuals would become so ferocious, and the difference between individuals so reduced, that the social structure would be destroyed. The cellular slime molds make the point even more forcefully. The stalk cells die as they become trapped within the stalk, and if they differed genetically from the spore cells, they would be selected out in one generation. The only way of maintaining the altruistic stalk cells in the population is by having range variation. This type of variation will be of special significance for organisms that develop asexually or clonally (including the case of the cell progeny of a fertilized egg in a multicellular organism). It is an effective means of keeping the population heterogeneous.

We know very little about the mechanism by which the range variation is achieved. With behavioral examples the difficulty is that the genetical basis remains largely conjectural. In using the term "noise," Waddington (1957) clearly shows that he considers this kind of variation in metazoans to amount to errors compounded in development. Perhaps this is correct for many of the examples, but one wonders if, in addition, there is not some kind of regulative control. In the case of social insects the controls are far more elaborate and limit not only the morphological extremes but the relative proportions of the numbers of individuals as well. Even in some protozoa, as in the example just mentioned showing bimodal size distribution upon starvation, there must be a method of regulation.

There is another aspect of range variation that leads rather directly into the next topic. In a cell system, along with variation in cell size and in the chemical composition of the cells, there is bound to be variation of the timing of certain cellular events. For instance, in the cellular slime molds, some cells ("founder cells" of Shaffer, 1961, in some species) begin the formation of the centers of the aggregates. There is considerable evidence that these cells are not genetically different from the cells they attract (review: Shaffer, 1962), and therefore we presume that they are genetically identical. The most reasonable hypothesis is that they vary phenotypically, and on the basis of chance the first to reach a certain stage begin forming a center. There is good evidence that once they begin this process they inhibit neighboring cells from becoming competing centers (Shaffer, 1963; Bonner and Hoffman, 1963). If all cells were identical in their timing, presumably there would be no possibility of aggregation, for each cell would be a founder. The organization is possible by virtue of the fact that the timing of the cell-changes varies, and we can reasonably assume that this is another instance of the significance of range variation in contributing to the pattern of development.

Differentiation by successive changes in time

One of the advantages of looking at organisms as life cycles is that a wholly new type of division of labor emerges. If, for instance, all the cells of an organism do one thing at one point in the life cycle and something else at another point, the labor may be divided temporally rather than spatially; it is in series rather than in parallel.

Some of the best examples come from multinucleate organisms. For instance, in the Foraminifera (Plate 4) and in the Myxomycetes (Plate 7) all the nuclei and the surrounding cytoplasm first go through a vegetative period of feeding and growth. This is followed by a reproductive period where the nuclei by

progressive cleavage are cut up into small sexual or asexual reproductive bodies. Therefore instead of some of the cells being
separated off as germ cells and the remainder as soma at one
moment in the life cycle, as is the rule for all higher animals and
plants, in these primitive forms all the nuclei and their associated
cytoplasm (or energids, to use Sachs' old term) first are vegetative and behave like the soma, and then the same energids
become germ plasm and reproduce. There are many other examples of temporal differentiation. Myxobacteria (Plate 5) do
precisely the same thing. So also do many of the coenocytic algae,
for instance *Bryopsis* (Plate 8), *Acetabularia*, or *Hydrodictyon*.

Furthermore, differentiation by successive changes during the
course of time is the rule in unicellular organisms. An amoeba,
for instance, may first feed and grow and subsequently go through
a period of encystment (Plate 2). The same is true of a sporulating bacterium, a flagellate (Plate 3), or even a ciliate. In the
life cycles of bacteria or other single cells having a series of
binary fissions the principle also applies. The cell goes through
a series of biochemical changes which bring the cell to and past
each division. They are, in fact, the sequence of steps, and at any
one moment the steps are different; in this sense there is a temporal differentiation. Embryologists and developmental biologists have frequently expressed concern with the difficulties of
analyzing developmental process by studying chemical control
mechanisms in bacteria. The answer may be that if we understand thoroughly the simple temporal differentiation which runs
in series, we may ultimately discover insights into the more complicated differentiation of higher multicellular forms which runs
in parallel.

There is an interesting point here concerning cultures or populations of unicellular organisms. Unless the cells are dividing
synchronously, a colony of loosely associated cells will contain
cells at different stages in their respective life cycles, and in this
sense the colony as a whole will show spatial differentiation.
However it is presumed that this differentiation is purely haphazard, and is in no way proportioned or controlled. If there is
any degree of control and a set pattern of differentiation is produced among the cells, we immediately jump from the level of
a loose population to that of a multicellular organism.

It is especially common to find cases where there is a combination of both spatial and temporal differentiation. For instance in the cellular slime molds all the cells are vegetative at first, but following feeding some of the cells go into the formation of the stalk while others go into the formation of the spore mass (Plate 6). In many fungi the situation is not quite so clear-cut. In some cases, as in many of the common molds (e.g. *Mucor, Neurospora*), all the nuclei may eventually flow into the sporangium, but in some parts of one large mycelium vegetative processes will be occurring while fruiting is simultaneously taking place in others. There is the interesting added point that during the vegetative phase the protoplast lays down a rigid cell wall that remains; the living substance can then move from this self-constructed tube up into the rising sporangium, leaving behind a dead but highly elaborate skeletal structure.

In a very general sense all organisms have some degree of temporal differentiation. For instance in mammals the kind of activity performed by an infant is different from that performed by an adult. This point is even more striking in organisms with a metamorphosis. In lepidopterans, for instance, all the major activities as well as the form of the caterpillar are very different from those of the adult. Recently F. C. Kafatos (1965) has shown that while certain cells in the caterpillar remain undifferentiated and indifferent until they reach the adult stage, others will first show a larval differentiation and then an adult differentiation. In the transition within the chrysalis the cytoplasm of these differentiated structures degenerates while the nuclei remain, and the new cytoplasm that regenerates takes on the new adult differentiation and function. As Kafatos points out, this illustrates the importance of considering organisms as life cycles, and also that all of the stages are functional in their own sense and not merely the adult stage.

Temporal differentiation also occurs in social insects. There is the well-known case in honey bees where the worker performs different functions in the hive depending upon his age (review: Lindaur, 1961). As can be seen from Figure 11, the worker during the period of twenty-four days graduates from janitorial work to nursing and construction work to foraging and dance-following. Since any one hive consists of a mixture of ages, there

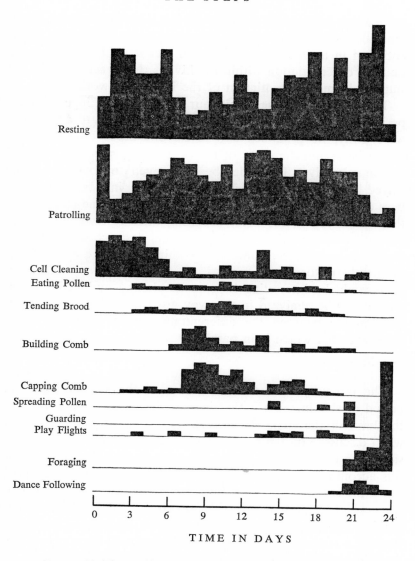

FIGURE 11. The total amount of work accomplished by a single bee during the course of her life. The work is classified according to type and plotted as per cent time spent on each occupation. There is clearly a sequence dependent upon age, passing from the phases of cell cleaning to brood care, to building, to guarding, and to foraging. Note the large amount of time spent in patrolling and in apparent inactivity. (After Lindaur, 1961)

are enough individuals to divide all the labor of the hive simultaneously; the individuals, on the other hand, show their differentiation in a serial time course. In fact it might be presumed that the differentiation within the hive is the direct result of the staggered time-differentiation of the individual worker bees.

To test this hypothesis Lindaur performed an ingenious experiment. By substituting hives when the mature foraging bees were away, he produced two hives, one made up entirely of young bees, and the other of the older foragers which returned to the new hive. The older bees managed to cope and some stayed back to do the nursing and housekeeping, but the young bees came close to a catastrophic end. At first none foraged but they all stayed in the hive, depleting their reserves and many even died of starvation. Just before total destruction a few bees finally set out to forage and, by this thin thread, the balance was ultimately restored.

It would be interesting if some developing multicellular systems could be analyzed in this way, for the method manages to discriminate between the differentiation which is in series in time, and that which is in parallel.

Conclusion

The difficulty with this discussion of the steps during the period of size increase is that it never has really come down to the level of the steps themselves. This is a vast subject, even when discussed on the general level used here; it involves a consideration of the period of development of all organisms, and if this account were to deal with the actual sequence of chemical steps, the story would be without end. Not only that, but it is a story which at present is largely unknown, and I believe that it is more important to try to find consistent patterns in the steps. Further attempts, again on a general level, will be made in this direction in the last section of this chapter.

We began with the notion that the problem of development should be framed by the question of how is it possible for a multicellular organism, with a large number of what we have assumed to be identical genomes, to produce a differentiated structure in which the labor is divided. It has been shown that sometimes the labor is simply not divided but only a crystal-like symmetry or polarity is produced. Sometimes, when the cells are rigidly fixed with relation to one another, chemical polar signaling in the form of gradients or inductions are employed. Sometimes, when the cells can move freely within the cell mass, there is a phenotypic variation among the cells ("range variation"); and the cells, on the basis of their haphazard differences, are selectively sorted out during organized mass cell movements or, more properly, morphogenetic movements. Finally, sometimes the differentiation takes place serially, so that at one moment all the units will perform one function, and then another at a later moment.

In closing this section, I want to stress again that it is unusual to have any one of these systems prevail alone. Frequently two or more will occur in a single organism, often at the same time; for instance, consider all the complexities of the development of an amphibian.

THE STEPS DURING SIZE DECREASE AND SIZE EQUILIBRIUM

The period of size decrease

In describing various types of size reduction distinctions were made between total and abrupt size reduction as in cell division or complete fragmentation into spores, partial and abrupt size reduction as found in spore formation and gamete production, and finally those cases of gradual size reduction like that of senescence.

There is an extensive literature on the subject of the nature of the stimuli which initiates cell division. This has recently been well reviewed by Mazia (1961), who points out that while critical size is the factor that provides the stimulus, the experimental data (particularly that of Prescott, 1955, 1956) shows that this can be so in only a very approximate way. Prescott demonstrated that by removing portions of the cytoplasm by microsurgery at regular intervals, division could be postponed for the entire duration of the experiment. Another significant series of experiments was carried out by Scherbaum and Zeuthen (1954) on *Tetrahymena*, where temperature shocks will block cleavage but not growth or DNA synthesis. When replaced under constant temperature conditions there is a spate of successive cell divisions which restores the original size relations. This resembles the normal situation in other ciliates and flagellates (Plate 3) and in other groups of protozoa (e.g. Foraminifera, Plate 4) where the whole cell normally breaks up into a number of small daughter cells (or gametes). In many of these instances, and this is an argument started by Hertwig (review: Fauré-Fremiet, 1925; Mazia, 1961), it is thought that the critical point may be tipped off by the nucleus-cytoplasmic size relations. The only real evidence, however, comes from the fact that in many of these cases there is a restoration of the nucleo-cytoplasmic ratios.

Once the signal has been given, wherever it may originally emanate from, the steps of mitosis and cleavage begin. As can be seen from Mazia's (1961) excellent summary of his own work and that of many others, there are already many known sequences of events concerned with cell division that take place within a cell. This can be said even though we know and understand what must be a very small fraction of all the steps. There are processes which prepare for and perform spindle formation, chromosome doubling, cleavage furrowing or cell plate formation, etc. Each of these events or chains of steps has a different timing within the cell cycle. It is even more interesting that these events can, by juggling the environmental conditions, be shifted in time or even eliminated without seriously affecting the other processes. Here again we see evidence, and in this case within one small cell, that the steps do not occur in a single linear sequence; instead there are groups or chains of steps running in parallel, and they are remarkably independent of one another.

Very little is known of the causes of splitting in multicellular animals, such as those species of planarian worms which normally pinch in two (Plate 20), but again it is presumed that critical size is the key. It should be remembered that in this case and in that of the cell there are really two problems; one is whether size is in fact the crux and the second is how a critical size, once reached, sends off a signal.

Colony splitting in social insects is better understood. In honey bees and army ants, where it has been studied closely, the prime stimulus is the appearance of one or more new queens.

But as with all internal stimuli, one then wants to know what stimulated the production of extra queens; is it merely a regular seasonal process genetically fixed, or is there some feedback, some relation to the size of the colony? As Schneirla and Brown (1950) have described in detail for army ants, once the new and the old queens are present, there is great confusion among the workers, for apparently they cannot serve two queens, which they recognize by their odor, and they must decide between them. The result is much like a rather frantic election campaign with much agitation and what appear to be violent partisan feelings. Ultimately two camps separate and go their independent ways. In bees the old queen leaves the hive, followed by a group of faithful workers who have become split off from the other workers who have riveted their attention on the new queen.

To come now to the very general case of partial and abrupt size decrease, let us first examine the question of spore formation in many lower forms (including a number of cases where spore formation is total). In the first place, sporulation is largely dependent upon a suitable environment; in particular it has been shown that a decrease in humidity, sometimes very slight, is sufficient to initiate spore formation. In some fungi it has been possible, as Morton (1961) has shown for *Penicillium*, to imitate this effect in submerged cultures by increasing the osmotic pressure of the liquid medium. But he also showed that a brief exposure to air is far more effective, suggesting that the air promotes a rapid change in the cell wall surface, leading to spore formation.

Given the proper temperature and humidity (and in some cases illumination as well), a particularly significant and common stimulus is starvation. This effect is general among fungi and occurs in both plasmodial and cellular slime molds. In the latter the separate amoebae will not aggregate if new food is constantly added, but in a culture that finishes its food supply, aggregation will begin after six or more hours. It is possible to show that this alteration from the feeding to the fruiting state is accompanied by morphological changes in the mitochondria (Takeuchi, 1960) and by staining differences such as an increase in metachromasia with toluidine blue (Bonner, Chiquoine, and Kolderie, 1955), but we do not know what these changes mean in terms of chemical steps.

Further progress has been made by Daniel and Rusch (reviews: Alexopoulos, 1963; Rusch, 1959) in the true slime molds or Myxomycetes. They have perfected a method of growing the plasmodia in liquid shake cultures with a simple medium that includes a small amount of embryo extract. In this method it is possible at any moment to centrifuge the plasmodium free of the growth medium, and to put it on fresh medium containing salts but lacking in organic substrates. In order for the plasmodium to produce fruiting bodies two requirements must be fulfilled: a period of illumination at the correct moment is necessary, as was known from the previous work of Gray (see Alexopoulos), and niacin or some suitable precursor of this substance must be present. By removing the plasmodium at various stages from these conditions and replacing them in the nutrient growth conditions, Rusch showed that there is a moment when the steps leading to sporulation cannot be reversed; they will proceed even if surrounded by sufficient nutriment. The many biochemical steps which lead to sporulation are still largely unknown, although the promise for future clarification is great.

Besides spore formation, another method of abrupt size decrease is gametogenesis. Fertilization is the initial step in size increase, and fertilization cannot occur until the eggs and sperm are ripe. In other words, the moment of initiation of size increase is totally involved with the moment of size decrease, for the same external factors promote gamete production and their suitable conditioning for fertilization. This is one more piece of evidence that the life cycle is a continuous process and can be fragmented only by abstract and artificial schemes of classification.

Let us delve into this particular transition by examining the sequence of events in a vertebrate, for example a bird. The initial timing signal is day-length, and once the spring days achieve a certain duration, the gonads are stimulated to a highly active period of growth. However, as Lehrman (1959, 1961) and others have shown, this stimulus affects the male more than the female, but the latter is in turn strongly influenced by the presence of the male. Ultimately, in the company of one another, the ripe gametes are formed. From this point on, as Lehrman has been able to discover in his experiments on doves, the sequence of events is a series of stimuli and responses, each of which is dependent upon a previous condition and is therefore truly a rigid sequence of steps. Besides the well-known steps of courtship, there are those of nest building, egg laying, chick rearing, and so forth. Each stage is utterly dependent upon the previous steps, and the process carries straight through from the growth of the gonads and the cutting off of the gametes to the rearing of the fledglings. It is of particular interest that this sequence not only involves the environment (the day-length) but also stimuli such as eggs or the presence and actions of the mate, and finally these are all coupled with the activities of the endocrine system. The external events affect the hormones that condition the physiological and behavioral responses of the individual, and this long complex chain, involving the two sexes and finally the offspring, brings the life cycle smoothly past the period of size decrease well into the period of size increase.

As a final example of size decrease, we come to senescence. Here, more than anywhere else, we are mostly in the dark concerning the nature of the steps. The difficulty may lie in the possibility that there is more than one way in which the decay can occur; the phenomenon is not identical for all organisms.

First of all, there are many known organisms in which there is no evidence whatsoever of senescence. Previously we mentioned the famous case of the sea anemones that were kept in captivity for many years (see Comfort, 1956), an example of an organism that maintains a constant size without decrease. Many large trees, such as the giant sequoia, grow continuously without apparent decline; their fall is finally a matter of destruction by the wind and the elements.

In some of those cases where there is a clear-cut senescence, it is thought that perhaps it is not a planned sequence of steps but an accumulation of errors, a breakdown of the system. This is the interpretation given by Sonneborn and his co-workers (see Sonneborn and Dippell, 1960) to clonal senescence in *Paramecium*. Since the macronucleus merely pinches in two automatically at every division, there are chances that some of the genes may fail to be in both daughter cells. But more important than this, there is evidence that the aging of the cytoplasm promotes somatic (or macronuclear) mutations which result in a general decline. Only some form of nuclear reorganization, that is the manufacture of a new macronucleus from the micronucleus, either after conjugation or autogamy, can produce a rejuvenation that affects the cytoplasm as well as the new nucleus.

In other instances of senescence there is apparently a well-organized series of steps that proceeds with predictability and precision. The case already cited is the regression cycle of the hydroids *Obelia* and *Campanularia* described by Crowell (1953), where the hydranths grow and then regress in definite waves.

Senescence in man is the most puzzling and complex problem of all. Perhaps part of the puzzlement arises because both of the above processes occur simultaneously in different parts of the body, and there is no one explanation for senescence in human beings. The problem is complex because all the parts of the body are interconnected in a complicated fashion, and finding single causes in such a vast interweaving meshwork is exceedingly difficult.

The period of size equilibrium

In senescence we saw that there could be a planned and an accidental sequence of steps, or both. This is equally true with the period of size equilibrium, for it can either be maintained by a rigorous series of steps or by total immobilization. The former would be a steady state (or dynamic equilibrium) while the latter is a static equilibrium.

In cases of total immobilization, such as spore formation, there are still the steps leading up to it, forming a strict sequence. But these are also the steps which lead to the abrupt size decrease, that is, the cutting off of the spores. Again we see that the steps of one period lead directly and continuously into the steps of the next.

Once the spores are formed (and this is true of seeds as well) they go into a kind of suspended animation. Equilibrium here is primarily achieved by doing nothing (i.e. static), or at least relatively little. Metabolism is extremely low, and as a result the longevity of the stage is greatly extended. This is only broken by the stimuli which lead to germination, but here we are also passing from one stage to the next, in this case into the steps that lead to size increase.

In those instances where there is no dormancy but an active organism, the size equilibrium is of a different nature. Here we are concerned with two problems: one is that of attaining a size limit, for there must be some mechanism which stops the organism from further growth; second, there must be some mechanism by which the size is maintained at a constant level with an active metabolism and synthesis of replacement protoplasm. We may presume that these two problems may be governed by the same mechanism despite the fact that stopping growth increase is somewhat different from maintaining an equilibrium. In any event, the critical size reached here is different from that reached in cases of cell division, for here, once the critical size is reached, it is maintained by a dynamic equilibrium, while in cell division, when the critical size is reached, it is immediately halved.

The mechanism of growth stoppage once an organism reaches a certain size is unfortunately poorly understood, even though there is great interest in the problem. The growth rate of an organism can be accurately described, like all its other measurable characteristics, but unfortunately this tells us little of the steps involved. We know that size limits are genetically determined and can be selected for by breeding. We know that in some dwarf or small races the small size is accompanied by a reduction of the rate of growth, while in others the growth rate is normal but the process stops sooner. In some old experiments on large and small rabbits Castle (1929) showed that in the large ones the rate of cell division was increased while the timing of the differentiation did not change, with the result that each equivalent embryological stage was larger and contained more cells. Furthermore, Sinnott (review: 1960) has shown that in large squashes and gourds there is a longer period of cell division and the cells are relatively larger. Perhaps the timing of stoppages is not so important as the fact that they do stop, while in some organisms they do not. It would be of great interest to know the exact nature of the steps (beginning from the genes) that cause this sharply defined end to growth. And also we should like to know the difference in the signals within an organism that is in the process of growing and one that is active yet constant in size.

There are a few steps that are known in some organisms. For instance, in mammals it is known that the endocrine glands, especially the growth hormone of the pituitary, are intimately connected with the ultimate size; an excess of growth hormone produces gigantism and a deficiency, dwarfism. There is even a genetic strain of dwarf mice in which the mutant gene produces a pituitary that is incapable of secreting sufficient growth hormone, and these mice can be cured of their hereditary dwarfism by injection of additional hormone. This, however, is only a fragment of the picture, for there are other hormones, such as thyroxin, that also affect the size of the individual. In addition, there is the question of what tissues are capable of growth and when. For instance, the diapho-epiphyseal junction of the long bones seals off eventually, preventing any further increase in height, and there is no evidence that this change in bone structure is caused by the pituitary.

In higher plants there is a similar situation. Auxin, the growth hormone, is clearly instrumental in size increase, and the ability to respond is dependent upon the competence of the meristematic tissues. In trees with a cambium and secondary thickening, the ability to expand continues indefinitely, or at least until weight-strength ratios become a problem, while in annual plants there may be a growth limit. The causes are known to be varied; not only are the auxins limiting in their action, but so are the giberellins and undoubtedly other substances as well. The complexities of the problem are all too frequently brought home to the plant physiologist.

Size limits and size maintenance are also of significance to societies and populations. The fact that the number of individuals within a population is maintained within a range has already been stressed. For many years the ecologists have discussed the causes of the regulation of animal numbers, and there are those who have emphasized the external, environmental factors (review: Browning, 1963) while others have championed internal regulatory mechanisms (review: Lack, 1954). In this latter category a very successful theory, at least for higher organisms, has recently been put forward by Wynne-Edwards (1962).

Briefly, he considers that a prime function of social relations among animal populations is the regulation of numbers. This is achieved largely by displays, and the bigger the gathering, and the more elaborate the system of conventional display, the more likely is it that an elaborate feedback system of population control has evolved. The display activities allow the individuals of a population to confront one another in a conventional pattern. If the population is low and food sources high, the individuals in the communal gathering will indicate the opportunities by enthusiastic and vigorous display activities, and the reverse is true in overcrowded conditions. The chain of events is that the pressures of population density affect display activities which in turn affect reproductive success.

To this bare outline he proceeds to add a large number of crucial details. For instance, the kinds of conventional displays are diverse; they may consist of territorial arguments between males, leks or the gathering of courting males found in some species of birds, the aggregations of animals such as starling flocks in the fall, or gnats, or a simple peck-order or order of dominance. Following Carr-Saunders he even suggests that the harvest gatherings of primitive societies serves this same feedback system of population control for man. Besides vigor in reproductive activity, there are other ways that the number of individuals in a population can be controlled: induced mortality (e.g. infanticide), deferment of maturity, and others. Furthermore, all of these variables have the additional advantages of flexibility, so as to be able, for instance, to replace large losses due to some climatic accident. The reader is urged to go directly to Wynne-Edwards' work for the permutations of his basic interpretation of the analysis of number control in animal populations. There undoubtedly are other factors that help to exert control, for one of the characteristics of control systems of this sort is to have more than one safety device. Nevertheless, by Wynne-Edwards' idea the regulation of numbers is accounted for, and a large share of previously puzzling social conventions suddenly take on meaning.

Since all these systems involve mobility and sense organs, they obviously cannot be present in plants. But some plants do have methods of density control, although not in the total numbers of individuals, which is largely determined by geographical factors. Each individual plant will produce an inhibitor that spreads over a given area and prevents any other member of the same species from germinating or growing within the circle.

To return to multicellular organisms, we have described the period of size increase, size decrease, and size equilibrium. It is now time to describe those steps that usually occur during the period of size equilibrium and that prepare the organism for the initiation of the period of size increase.

To begin, it will be recalled that small organisms are in general more exposed to the environment and more subject to environmental stimuli than large organisms. Since we are dealing with the transition between the period of size equilibrium and size increase, the stage involved is generally a single cell or at least a small stage in the cycle: a spore, a gamete, or a seed.

For example, the germination of spores and seeds of plants is directly dependent upon environmental stimuli. The proper temperature, humidity, and chemical or nutrient environment are required before initiation will take place. As a rule these conditions are also the ones which are favorable for growth, for in this way germination can be directly followed by continued development. Obviously, this is the adaptive advantage of the environmental trigger.

There are some interesting exceptions to this rule. For instance, in certain desert plants ideal growth conditions will not promote germination. Instead, a considerable rainfall is first necessary in order to leach out substances that inhibit germination and which lie in the seed coat. In this discussion we are continually encroaching on the subject of the next chapter, but we must point out here that this adaptation has the advantage of allowing the seeds not only immediate suitable growth conditions, but prolonged ones so that the next generation of seeds may be reached before the water gives out.

In many cases the germination is dependent upon environmental stimuli in more complicated ways. In particular, the seeds of many plants, the gemmules of sponges, the pupae of numerous insects, and many other organisms require a period of cold before the growth conditions appear in order for initiation to take place. Often just subjecting the resting stage to the growing environment without previous chilling will result in a failure to germinate. For instance, in some sponges it can be shown that the length of time in cold pre-treatment is (within a given range) inversely proportionate to the time required for germination (Rasmont, 1962, see Fig. 12,a). Again the advantage to the organism is obvious, for in this way it is assured of germinating only in the spring when the future prospects of summer are warm and bright, and not in Indian summer when the oncoming bleak winter is sure to kill off any exposed growing tissues. By contrast, there are other species of sponges which do not require such vernalization (Fig. 12,b), and in fact Rasmont has shown that one species carries both types of gemmules. Each case presumably has different ecological advantages, for both exist in the same environment.

In some plants the germination process is not completely initiated by external stimuli but is strongly influenced by neighboring spores or seeds. There are instances where a seed that has successfully germinated, or a plant already growing, will stop the germination of seeds of the same species that lie within a given radius by giving off an inhibitor substance. Among many fungi and the cellular slime molds the per cent germination is greatly reduced by concentrating the spores (Russell and Bonner, 1960, see Fig. 13,a). This is advantageous since any one spore can start a new generation, and therefore the suppressed, ungerminated spores can be used again. This provides an interesting contrast with the case of Myxomycetes, where Smart (1937) showed that for some species the per cent germination was increased when germinated in water taken from previous germinating spores (Fig. 13,b). Here the products of germination are gametes and, therefore, clearly a more crowded population of swarmers would increase the chances of fertilization.

In the eggs of animals and plants the initiation mechanism is largely by an internal stimulus. That is, if a sperm fertilizes an egg, activation usually ensues, and in a sense this must be considered internal. It must be remembered, however, that it is possible in cases of parthenogenesis, artificial and natural, to activate the egg by environmental cues without the help of the sperm.

The process of fertilization itself involves a beautiful series of steps which are still in the process of being elucidated. For the

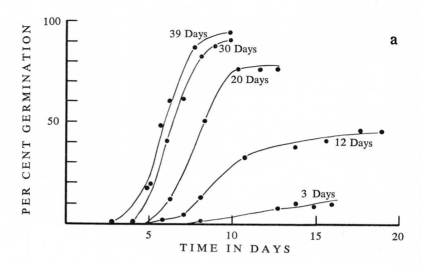

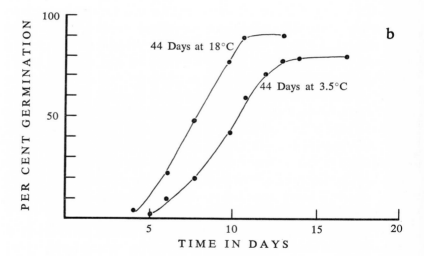

FIGURE 12. The effect of cold on the germination rates of sponge gemmules. (*a*) The germination rates for *Ephydatia mülleri* at 20°C after being previously incubated for varying periods at 3.5°C. (*b*) The gemmules of *Ephydatia fluviatilis* were incubated at 16°C after two different pre-treatments. Note that in *a* the cold pre-treatment is decidedly advantageous while in *b* it is not. (From Rasmont, 1962)

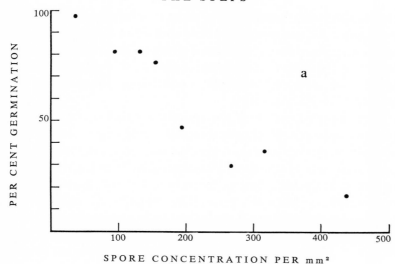

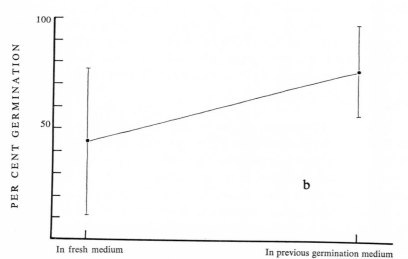

FIGURE 13. The effect of cell density on spore germination in different types of slime molds. (*a*) The per cent germination is plotted against the spore concentration for the cellular slime mold *Dictyostelium mucoroides* (from Russell and Bonner, 1960). (*b*) Fifteen different species of true slime molds (Myxomycete) were tested by placing 10 spores each in fresh medium and in medium in which previous germination had occurred. The means and the standard deviation for the 15 species are indicated. It is clear that in the majority of species there is an advantage to germinating in the old medium, but this is not true for all (data from Smart, 1937).

entry of the sperm into the egg a series of substances is produced in sequence by one gamete, which stimulates the other to produce a specific substance which in turn affects its partner. By such highly specific chemical conversations it is possible for the egg and the sperm of the same species to combine at the correct point in their maturity.

Even though these steps are entirely internal in that they involve the passage of chemical signals only from cell to cell, there are important external signals also, even for those higher forms that have internal fertilization. The external signals are concerned again with timing, for it is important that both gametes mature at the same time. The majority of animals and plants are not fertile at all times, but only at certain seasons of the year. It is at those moments that the rut takes place, or flowering in the case of plants, and this obviously must happen for both sexes at the same time. As is well known, the important external cue is day length, the photoperiod, and by this universal means sexual ripeness can be synchronized. Response to the day length is as important to plants as it is to animals; both use the same system to time their seasonal fertilization. Therefore, even fertilization depends to some extent on external signals.

Conclusion

From this discussion we are struck by the fact that the steps of one stage lead directly to the steps of the next. The life cycle is one continuous process, and this point again emphasizes that our arbitrary subdivision of the cycle into different periods based on size is in some respects artificial and analytical.

The size yardstick is convenient, however; for instance we can say that in general those periods in the life cycle when the organisms are small are relatively more influenced by external stimuli than those stages that are large. This is consistent with the notion that in general size and complexity (in our sense of differentiation) tend to make the organism less dependent upon the vicissitudes of the environment. This is helpful in comparing adults of different organisms or one organism at different points in its life cycle.

We had problems of identifying all the steps during the period of size increase, and the number of steps and the magnitude of the problem become even more staggering if we consider the whole cycle. In the final section of this chapter we shall again see whether, having surveyed all the parts of the life cycle, we cannot make some further significant generalizations on the characteristics of the steps.

THE GROUPING OF STEPS INTO UNITS AND THEIR DISSOCIABILITY

Heterochrony

Genes act at all moments of the life cycle and not just in the period of development or size increase. This well-known fact has two implications of importance to us. The first is that it is the life cycle as a whole that is the significant unit (despite the problems of great interest in the period of size increase), and each step of the cycle (and therefore a gene action governing any part of the cycle) is as important as any other to keep the cycle going from one generation to the next; there can be no weak link in the chain. The second implication is that through mutation and selection any part of the cycle can change, and the whole structure or sequence of the cycle may become altered, as is implied in the term heterochrony.

The phenomenon of heterochrony has been amply discussed by de Beer (1958) in his significant book *Embryos and Ancestors*. He points out that Goldschmidt (1938) showed that genes determine the rates of processes, and that by gene mutation these rates can be changed with the result that certain stages in the life cycle can be contracted or protracted. With this idea as the basis of heterochrony, de Beer proceeds to give many examples from animal life histories where the timing of the events in the ontogeny has been altered during the course of phylogeny. He classifies these alterations in rather complex categories; for our purpose some simplification is appropriate.

There are two basic types of heterochrony: one is the loss or the addition of some step (or series of steps) in the life cycle, and the other is the shifting in the timing of any event within the life history, either forward or backward with relation to other events. (Strictly speaking only the latter fits under the meaning of the word "hetero-chrony" or altered timing, but it is convenient to include both under the one name.)

Additions or losses may be sudden or gradual. The whole sexual phase may be lost from the life cycle of some organism, and subsequently it reproduces only asexually. This would represent a sudden alteration, while the appearance and slow increase in importance of some character during the larval stage or adult stage would be a more gradual process since the trait keeps changing with subsequent selection. However, the distinction between sudden and gradual is merely one of degree and is not important.

Shifts in timing are very common; the best known example is neoteny. Here certain characters associated with the larval form appear progressively later during the period of size increase, until ultimately an organism bearing ripe gonads still possesses larval or immature physical characters. This could also have arisen by the progressive shift of the ripening of the gonads to an earlier period (paedogenesis); however, it is virtually impossible in any particular case to know what has shifted. In fact both the larval traits and the point of gonad maturity may have changed, and the only important matter is the relation of their timing to each other.

It is possible to take all the categories of de Beer and put them into the kind of classification we are suggesting here (Fig. 14, which uses de Beer's system of notation). But if one looks at actual examples of shifts in timing, it is often difficult for a particular case to place it in any one of the sub-categories with any certainty. The reason is that the timing can shift in different ways, and these ways may intermingle and overlap. Therefore the fine distinctions between categories are of little significance compared with the central fact that shifts in timing can and do occur.

Chains of steps

If we now return to the question of the sequence of steps, the phenomenon of heterochrony gives us some important clues. In the case of additions and losses, it is clearly possible to remove or interpolate a sequence of steps within a life cycle. This means that we must be cautious about the statement that in a life cycle one step leads to another in a rigid sequence. If we can remove or add without obliterating the cycle, then clearly each step is

ADDITIONS (OR LOSSES)

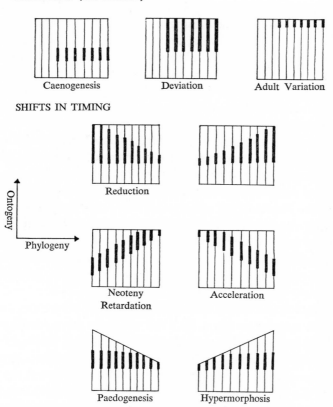

FIGURE 14. De Beer's (1958) system of notation for the various manifestations of heterochrony. These have been re-grouped in such a way as to stress the major kinds of changes in timing. Each line represents the development of an individual. The successive lines from left to right represent phylogenetic progression. The heavy bands on the lines indicate the presence of a particular character during a specific period of ontogeny. New characters can appear (or disappear) suddenly, or they can gradually shift their timing during ontogeny.

not completely dependent upon the previous step. Undoubtedly there are certain steps or series of steps that, if they were removed or tampered with, would cause the cycle to stop. But there are also obviously some steps or series of steps that can be removed as a unit or added by interpolation, without any catastrophe.

In the alteration of timing there is a similar lesson to be learned. In the first place, it should be stressed that there are two ways in which the timing can be altered. The first is the time at which the gene acts, and the second is the rate at which the gene action occurs. By either means, or by both, it is possible to produce shifts in the timing of two or more processes in relation to each other. For instance, the appearance of ripe gonads can shift with relation to the period of larval gills in amphibians so that neotenous forms are produced such as *Axolotl* or *Necturus*.

There are cases, such as one cited by de Beer, where the order of two processes becomes entirely reversed. Teeth are phylogenetically more primitive structures than a tongue, yet in mammals the tongue develops after the teeth. Or consider for instance the heart of the chick, which appears long before any of the other organs. The reason is undoubtedly connected with the necessity to transport food and gases for the proper metabolism of the embryo, but the fact remains that one structure appears and develops quite out of order with respect to the other structures.

If these cases are considered in terms of sequences of steps, it would appear to mean that the chain of steps leading to the construction of one of these structures, such as the heart, is not necessarily dependent upon a set relation to other structures. The formation of the tongue does not affect or control the formation of the teeth, for one can either precede or follow the other. It would seem rather that each process is a set unit or a series of relatively rigid sequences of steps that is independent of the other. Thus two or more of these chains of steps can be running side by side, with the result that the development of any complex organism may be made up of a series of loosely connected epigenetic canals (to borrow Waddington's term) running parallel to one another. Add to this the fact that whole sections of the cycle can be removed or added by interpolation (such as the loss of the sexual phase in a life cycle), then it would appear that the chains of steps can occur both in parallel or in series, and it is possible to shift these units in relation to one another without destroying the fabric of the cycle. This is a very similar situation to that described previously for the events leading to mitosis; because as Mazia (1961) points out, there is a series of chains of chemical steps affecting different aspects of cleavage and mitosis, and each is individually modifiable by experiment without affecting the other chains.

However, this does not mean that in the whole life cycle each of the units or chains does not have some relation to the others. Regardless of when the heart or the tongue appears in the cycle, a certain set of conditions must stimulate its appearance. It must be true that once certain chains are set into motion they will continue on their own in a self-contained fashion, but they still must be initiated. Therefore in heterochrony we can assume that the initiating stimulus for a chain of steps must have shifted in time, or there must have been a shift of the ability to respond.

Variation and the chains of steps

Thus we see that the life cycle consists of a large number of fairly rigid chains of steps which are hooked up in series or in parallel, and from the facts of heterochrony it is evident that the chains are dissociable as units. It is possible not only to shift the time of appearance of one of these chains by mutation, but it can be eliminated entirely or interpolated *de novo* in a cycle. Mutation can also, of course, make changes within a chain of steps, but in these cases the changes do not involve heterochrony but the chain itself, such as the structure of the tongue or the teeth.

If the steps did not occur in blocks or units that can be shifted or altered *in toto* without seriously affecting the rest of the organism, evolutionary change in complex organisms might have been virtually impossible. It would have meant that a slight alteration in one small link might have completely disrupted the whole chain, for each step of each part would be coupled into all the steps of neighboring parts. But if there can be separate autonomous units, then all that is presumably needed is to alter the initial cue or stimulus which initiates it; if the cue is moved forward in time relative to the other processes, then the unit as a whole will move forward. If there is a genetic suppression of the cue, then the whole unit will be eliminated. This, of course, could occur for other reasons also, but the consequences to the rest of the organism of this surgery would be minimal because the unit was independent. Other units may be dependent upon the eliminated unit for their own initiating stimuli, which would mean that elimination would have to be accompanied by more than one mutation. Perhaps in the future of experimental embryology and development in general, the isolation and identification of the chains of steps for any one organism may be of great importance in the analysis of development.

It must be added that it may not always be possible to detach or shift the units. De Beer gives some examples where a series of stages is preserved even though only the final one is functional. A case in point is the apparent recapitulation in the development of the avian kidney, where the stages of ammonia secretion and urea secretion are passed through before reaching uric acid secretion. Of course it is possible that, useless though these steps may be, they have simply failed to be eliminated; but it seems more probable that they are part of a rigid chain and cannot be eliminated piecemeal. These points can be settled only by experimental analysis of each step, but at the moment there is a clear suggestion that some chains are more closely knit than others.

The degree of dissociability of any one chain is important for the consideration of both evolution and the mechanism of the steps within the life cycle. Long ago von Baer pointed out that variation more commonly occurs at the later stages of development by a progressive deviation; in the earlier stages resemblances between embryos of widely divergent animals are striking, even though the adults are totally different. This well-established and accepted "law" of von Baer is pertinent here and can be expressed in terms of chains of steps.

What it says is that chains of steps which occur near the point of maximum size of the life cycle are far more likely to show variation than ones that occur near the point of minimum size. It is relatively easy to dissociate the chains in various ways when the organism is rapidly expanding in size. Assuming that dissociability can be represented by the degree of separation of spindles (which represent chains), Figure 15a shows the life cycle and the progressive degree of dissociability.

We are now faced with the interesting situation found in a number of organisms where there is a violent metamorphosis which separates the life cycle into two periods of size increase

(Plates 22, 23, 24, 25). For example, in insects (Plate 22) there is a considerable reduction in size as the caterpillar transforms into a butterfly; the imaginal discs really begin an embryonic process all over again during the process of pupation.

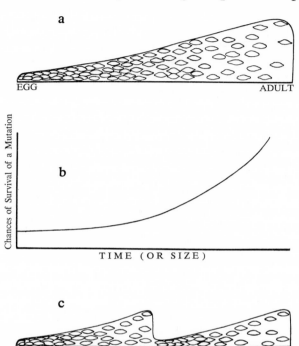

FIGURE 15. (*a*) Each spindle represents a chain of steps, and the purpose of this diagram is to illustrate the fact that the steps are more freely dissociable as the size within the life cycle increases. (*b*) A hypothetical curve showing the chance of survival of a mutation during the course of the life cycle of an organism in *a*. The longer the cycle and the larger the organism, the greater the chances of survival of a mutation if its effect appears late in the cycle. (*c*) The same for an organism which undergoes a complete metamorphosis, showing that there can be two such periods of increasing dissociability, one in the larval stage, and one in the adult stage.

Therefore it is not surprising, as de Beer (1958) has pointed out in some detail, to find many insects in which the larval forms show even greater variation between two related species than do the imagoes (Fig. 15,c). By metamorphosis it has been possible for the organism to have two periods of size increase and correspondingly two periods of greater variability or dissociability of the chains.

This double period of dissociability exists only where metamorphosis represents a genuine reduction of the central part of the organism, as in insects (Plate 22) and echinoderms (Plate 23). However in ascidians (Plate 24) and amphibians (Plate 25) the principal size reduction is the loss of a tail; the body itself undergoes a steady size increase from egg to adult. Another good example of a genuine size reduction occurs in the colonial hydroids (Plate 27), and here there is also ample evidence for variability in both the polyp stage and the medusa stage.

It is often said that variation is more prevalent in later stages of development than in earlier stages because mutations occurring in early stages are very likely to be lethal. The late-appearing mutations, on the other hand, are more likely to be superficial and therefore viable. To put this idea in the terms used here, alterations of steps or chains of steps are more likely to be lethal if they occur when the organism is small rather than large. It is of course a matter of probabilities, and spore color of a fungus could be no more lethal than eye color of a mammal, but often a gene effect has pleiotropic ramifications, and therefore even with this apparently innocent character there might be deleterious side-effects that would be more dangerous to the spore than the growing mammal. Figure 15b represents this idea in a hypothetical or diagrammatic graph. If the chance of survival of a mutation is plotted against size in the life cycle (which is correlated with time), the curve rises as the size increases.

A new idea can now be added to this scheme. We can compare an organism which has a long period of size increase and achieves a large size, such as a tree or a mammal, with a unicellular organism. If this is done, then on the same curve (Fig. 15c) the unicellular organism will occupy only a small fraction of the curve at its very beginning. This means that any mutation at any stage of the life cycle of a unicellular form has much less chance of survival than a mutation that occurs during most of the cycle of a higher form. In the case of the unicellular form, the loss is compensated by the rapid multiplication of other successful individuals, and if the mutation rate is sufficient, evolutionary change can continue to occur. We cannot say from this kind of speculation whether there are more mutations in one of our examples than in another; we can only comment on the probability of survival if mutation occurs, and how this probably is correlated with the size of the organism (and therefore complexity in the sense of differentiation).

Conclusion

In this chapter we have discussed the nature of the steps, but because of the magnitude of the topic we could do no more than try to discover some useful generalizations, briefly summarized here.

1. The cell, including the chromosomes, the remainder of the nucleus, and the surrounding cytoplasm with all its attendant structure, is the minimum unit of inheritance that joins one life cycle to the next. The point of minimum size in the cycle is therefore also the smallest possible unit of heredity.

2. The steps can be affected by either environmental or internal stimuli; in many cases the steps may respond to both. With increased size there is an increase in insulation from harmful or unwanted environmental stimuli.

3. The steps during size increase are of special interest, particularly because in multicellular organisms the genome of each cell may be assumed to be identical, yet differentiation results. There are a number of possible solutions of this apparent difficulty, and these solutions are to a considerable extent affected by the type of cell wall (and hence mobility) of the cells involved.

4. An examination of the steps during other phases of the life cycle shows that the steps lead from one phase to the next without interruption. The preparations for any one period can be traced well into the previous period.

5. The steps, from the gene level on up, can affect one another (e.g. pleiotropy); the chains of steps can have mutual controls one upon another.

6. Despite this criss-crossing of internal effects, many of the steps are in units or chains, and a whole chain can be moved forward or backward in time in the life cycle, or even eliminated completely, without necessarily having devastating effects

upon the other steps of the cycle. This bunching of steps into units is a tremendous help in producing viable variations in large multicellular organisms, so necessary for evolutionary change. Steps, therefore, are not only in series and in parallel in the organism, but in units or bundles also. Perhaps it is of even greater importance for experimental biology to understand the architecture of this network of steps than to try at this moment to identify the chemical events of all the steps. However, it is quite conceivable that our understanding will come through an essentially molecular or biochemical approach to the problem.

5. Evolution

The nonselectable

IT is pretty strong language, in this age of Darwin, to say that anything in a living system is above the force of natural selection. It would perhaps be more suitable to qualify by making it clear that what we are calling nonselectable is that which has not been affected by selection since the precambrian; we presume it must have been subject to natural selection before that. In other words, we are referring to those constituents of living organisms that are invariant or ubiquitous. It could be argued that they exist in all organisms because they are maintained by selection. They nevertheless are so indispensable to all living organisms that this becomes almost a trivial statement; it is like announcing that an automobile cannot run without a motor.

If we were to neglect momentarily the bacteria and blue-green algae and keep with what Picken (1960) calls eucells, then we can identify a number of invariant structures. For instance, there are nuclei enclosed in proper nuclear membranes containing conventional chromosomes. There are mitochondria, centrioles, spindle fibers, golgi apparatus, endoplasmic reticulum, ribosomes, etc. This is simply the standard cytologists' catalogue of cell parts. These structures are present in all eucells, and therefore it is assumed that their presence is essential for the existence of the cell. They are, in fact, the minimum requirement of a eucell, and thus we return to the notion of minimum size in the life cycle which was previously coupled with the minimum unit of inheritance. The cell is not only the irreducible unit of heredity but also the irreducible unit of metabolic maintenance.

While there is every probability that eucells have existed since the precambrian and have remained constant in their character, bacterial cells are assumed to be even more primitive and to have been the precursors of eucells. It is impossible to know whether or not this assumption is correct, but there are so many common features between the two types of cells that it seems reasonable that they should be phylogenetically related.

Their main difference is their size: bacterial cells are smaller than eucells. The change of size in evolving from a bacterial cell to a eucell probably produced a number of specializations or differentiations required by shifts in surface-volume ratios and the principle of similitude. Here is another instance of the principle of magnitude and division of labor. For example, many enzymes found on the surface of bacteria are found on the convoluted surface of the mitochondria in eucells.

Therefore, if we wish to find common denominators in all cells including bacteria (and the peculiar blue-green algae), we would not choose all the cytological components of eucells because many of them are absent in the smaller bacterial cells. However the biochemist has known for some time that the chemical constituents of bacterial cells and eucells are remarkably similar. Not only are the nucleic acid systems the same but also all the major enzyme-metabolic pathways. This is the basis for the often-heard remark that bacteria are as complicated as higher animals and plants, for they have the same Krebs cycle and other numerous similar chains of biochemical steps. (And of course in a biochemical sense this is correct, but in this book we are using complexity only in the sense of division of labor and differentiation.)

This means that to make a truly general statement as to what is rigid and invariant in all cells of all types, we should leave the realm of cytology for the realm of biochemistry. We see that there are a number of possible ways that biochemical steps can be structurally arranged (e.g. bacterial *vs.* eucell structure) but that the pattern of reaction pathways is the same for all. This, then, is the paramount invariant and irreducible aspect of living systems, so much so that we may refer to it as nonselectable. At least it is above selection *de facto*, for it is too much at the heart of the working machinery of life. It is a system that works, and apparently no serious tampering with it has occurred, at least since the precambrian.

The selectable

Besides the basic core of biochemical activity, there are changes in the chemical steps that affect the periods of size increase and size decrease in the life cycle. They clearly are altered by natural selection; the evidence is in the great variety of life cycles that have appeared since the first signs of life on earth. A great multitude of changes have occurred, and there are apparently many ways in which the biochemical steps in the life cycle can be altered without total destruction of the cycle, although this is of course not the case for the "nonselectable" steps. We should realize, however, that in this assertion we are making a gross assumption about the biochemistry of primeval cells merely on the basis of their morphological resemblance to present-day cells.

To turn to the question of what kinds of changes have been made, or better, where in the sequence of steps, selection can act, it is immediately apparent that selection can operate upon any of the steps. To illustrate, we may use again the system of notation developed in the chapter on method, adding a series of vertical arrows to indicate where selection can act.

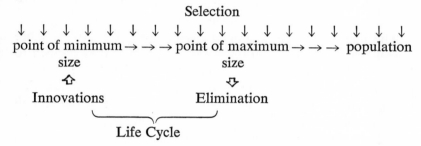

From this we see that selection can occur at any point in the life cycle and even beyond at the level of the population. Furthermore it shows that these changes can only be introduced at the point of minimum size, and they are eliminated at the point of maximum size. In other words, despite the fact that the points of introduction and elimination of changes are restricted, the changes themselves can occur at any point in the life cycle and even have effects that lie beyond at the population level.

This brings forth an old question: are all phenotypic characters adaptive? The question can be put in even more general terms: are all the steps in the life cycle adaptive? Since all of them can be altered by selection (except for the essential irreducible ones we have already discussed) then theoretically all of them could be adaptive, but if we analyze this sweeping generalization, it appears most improbable.

A step newly introduced by chance mutation could be selectively neutral. If this were the case there would be no selection pressure to eliminate and therefore it could conceivably remain. By a further mutation it could lead to another step that was adaptive, and since it has now become a necessary step in a chain of steps, it will be retained even though it is adaptively neutral. A character could be neutral for another reason also. In an ancestor there may have been a step which was adaptive, but because of changes in the environment it is no longer so, but merely neutral. Because of its neutrality there has again been no pressure to remove it; it remains because of neglect.

In order to test some of these ideas against actual examples, we might begin with the old question whether only the adult is adaptive, and the developmental stages are simply preparatory for this great moment. This view would be hard to defend, for although there may be inadaptive periods of development, they must be short and less conspicuous than the adaptive ones.

The matter can be examined where there are two or more phases which seem totally unrelated except that they happen to be part of the life cycle of the same organism. The simplest cases are those of organisms with a larval stage or an alternation of generations. In some, the adaptive significance of the second stage is obvious. In ferns the prothallus apparently serves the function of permitting egg and sperm to unite, although it is a device and a stage that has all but disappeared in the gymnosperm and angiosperm descendants (Plates 16, 17, and 18). In ascidians, the motile larvae may disperse, while the immobile adults cannot (Plate 24).

This list could be greatly expanded with a multitude of well-known examples, but of even more interest are those cases where the selective advantages are obscure and puzzling. For instance, one may wonder why the wheat rust spends part of its cycle upon wheat and part of it on barberry. One may wonder why the liver fluke has such an elaborate phase inside snails as well as sheep. Or to turn to nonparasitic examples, there are fantastically complex metamorphic changes of echinoderms (Plate 23), insects (Plate 22), and many other invertebrates which may mean a totally separate and elaborate independent existence of two stages that seem only barely continuous because the metamorphic changes are so violent and radical. Why, in terms of efficiency, is it not more reasonable to have echinoderm and insect larvae produce egg and sperm, rather than undergo the wasteful and complicated metamorphosis and start all over again as a virtually new individual? Would it not be simpler to have two species for each insect, one which inhabits the environment of the larva and the other that of the adult?

These problems and questions may have a familiar ring, and in fact it is largely due to Garstang (1928) and later de Beer (1958, see also Costello, 1961) that serious thought was given to the matter of larval adaptations. It is clear that all the structures and functions of larvae are undoubtedly adaptive and are as much objects of selection as adult stages. Furthermore the adult stages are sometimes eliminated or reduced by neoteny, and from this Garstang made his famous suggestion that because of the similarity of the larval forms of echinoderms and the hemichordate *Balanoglossus,* the chordates had arisen from the echinoderms by neoteny.

But the fact that many of the stages of a larva or a gametophyte in a plant are adaptive does not really explain why they exist at all, or why they arose in the first place.

Since the event occurred millions of years ago it is impossible to reconstruct, but we do know one thing with certainty. Many steps must exist if a big organism is to be produced, for large size is not achieved by instantaneous magic. In the evolution of large organisms there has been an increase, an addition of the number of steps. This has occurred by selection, and in the course of time in the changing world, some of the steps have remained adaptive, some have lost their adaptiveness, and some conceivably never had it. All that is really needed is that the total balance sheet of the life cycle favor the adaptive steps.

So far, these examples have been hypothetical rather than demonstrated. There are some genetic reasons, based on firmer evidence, that all the steps need not be adaptive. Since almost all gene actions have more than one effect, it can be reasonably assumed that some of the effects are more adaptive than others. A change can persist not only if some of the effects are neutral but if some are inadaptive, provided the other effects are strongly adaptive; the over-all balance must be positive. Furthermore, as Mayr (1963) discusses in some detail, modifying genes can alter some of the deleterious effects so that they are suppressed or become masked. The result is that the concept of the adaptability of any particular step becomes somewhat unreal, for each step is connected to a meshwork of steps; each change affects many others, and the adaptive value of each is related to the adaptive value of all the other steps to which it is connected, even those that are remote. Every time we try to pin a property to a single step we again realize that the steps are in a complex pattern, and it is almost useless to talk about them in isolation. This brings us back to the point that the understanding of the pattern of steps within the cycle is the great objective of modern experimental biology.

Selection at the population level

Natural selection affects populations in a number of ways. The most obvious is that the process of elimination of individuals and the manufacture of new individuals can be accomplished only in populations; selection is meaningless with single individuals. The more important point is that certain characters or individuals that are strictly social serve to bind a society in varying degrees of integration. These characters are useless for lone individuals, but can come into play only when a number of individuals are together. They consist primarily of communication systems among individuals; they are a set of signals and responses to signals that bind the group together so that concerted action in feeding, mutual protection, or some other group function can be effectively achieved.

These communication systems have adaptive value. For instance, if a group of birds or monkeys can give an alarm note and respond to it by fleeing or hiding, this will obviously reduce the loss by predation and therefore be adaptive. The same would be true of cooperative hunting in wolf packs, where a group of wolves together can be more effective than a single wolf in running down the prey.

The ability to give off the signal and the capacity to respond in the correct way are controlled by the genetic constitution of the individual animal. This means that these particular kinds of innovation were introduced (as were all others) at the point of minimum size of the life cycle of one individual, and selection of them occurred at the point of maximum size. The population itself, even though it is subject to selection, cannot innovate or eliminate; this can only be done at the level of the life cycle of the individual. Therefore all changes which affect the social integration of populations do so by extrapolation from the life cycle. For years there has been considerable discussion whether all social or altruistic traits could evolve by group selection, that is by all the individuals in one population being successful while a competing population was eliminated *en masse*. The difficulty is that what has in fact occurred during evolution is unknown, and one can make rational arguments that social characters arose by group selection or by individual selection. Perhaps both occurred, but in either case the genes must have been introduced or eliminated in the individual life cycles.

There are other aspects of populations that are of considerable interest, especially when one attempts to see how their properties differ from those of life cycles.

Again, questions of size are pertinent. In a population "size" means the number of individuals, and for the moment let us consider multicellular animals, which have been most extensively studied by animal ecologists.

A life cycle fluctuates regularly from a set minimum to a set maximum size. As has already been discussed, there is mounting evidence that a population, as nearly as possible, maintains its size at a constant level. Certain species undergo cyclic fluctuations in their population size, but even here there is evidence that the size is kept within limits. Too few individuals will result in extinction, as would too many; homeostasis is of prime importance and therefore it is in itself an adaptive character.

In other words homeostatic mechanisms themselves are subject to selection, and recently there has been great interest in the nature of the mechanisms. We have already discussed Wynne-Edwards' (1962) suggestion of a number of ways in which animal behavior, such as aggregations of individuals, sexual displays, and territoriality, could serve this function. All of these are characters which are genetically circumscribed and therefore handled by the mechanism of the life cycle.

Homeostasis of population is influenced by another important factor, as has been shown by MacArthur and by Hutchinson (review: Hutchinson, 1959). If there are many different species of animals in any one community, stability will result, and the populations of each species will be controlled in their fluctuations. In this case the advantage comes entirely from the complexity of the community. It is akin to the stability we observed in larger complex multicellular organisms. Complexity brings stability, and therefore complexity in itself is adaptive, both in the community and in the individual organism.

But we have now finally come to a character that is not controlled by the genome of one species: the complexity of a community is dependent upon the genomes of many species. Since multiplying the number of species is advantageous, the question arises as to the mechanism of species formation.

We are fortunate that this important and interesting question has been examined in detail by Mayr (1963) recently. There are, of course, many ways in which breeding isolation can occur between two populations that lead ultimately to species formation. It is easy to make conjectures about the sequence of events, but because the process, no matter how it occurs, is bound to be in some respects gradual and continuous, the question of how taxonomically one describes a species remains a messy and controversial subject which is not likely to be cleared up by any number of flat, arbitrary assertions. The important points are that there are discontinuities and that there are constant pressures to produce discontinuities, not only for the advantage derived from community stability but for the whole basic system of natural selection. Evolution cannot make steps forward if, by interbreeding, there is constant slipping backward. Reproductive barriers are necessary to keep change from dissolving. But because discontinuities can arise gradually and continue smoothly and progressively, systematics will always be painful for those who seek a rigid order.

Some of the causes of discontinuity are environmental, and geographic barriers present the prime example. But many are genetically derived and therefore bring us back to life cycles. Behavioral differences may prevent mating between two populations living in the same region, and they may involve the time of mating activity or the ritual of courtship. There may be direct genetic effects where cross-mating is not initially inhibited, but the offspring are inviable and this first step may lead to further differences. In each of these cases the initial steps come from mutation and selection; they are introduced in the standard fashion into the life cycle. Once they are there, through the same system of the life cycle, the two populations can build further changes that lead to evolutionary change. It is often said that the species is the unit of evolution and in a sense this is so. But so are the subspecies, the variety, and the genus, if we could clearly mark out where one stops and the other begins. In a much more satisfactory sense, the life cycle is the unit of evolution, for here the innovations are introduced and eliminated. And here there is no discontinuity; the life cycle can be defined as a unit, with sufficient precision to be satisfying to all.

Conclusion

In our discussion of the relation of evolution to the steps, we began by considering those steps that are invariant and appear, at least for millions of years, to be above and beyond selection. Those are the biochemical steps involved in basic cell metabolism, which are set aside in a characteristic structure in the small cells of bacteria, and a different and more elaborate structure in the larger eucells. This minimum invariant core is also the minimum point in the life cycle. Natural selection operates on the steps which lead from this set of core steps; that is, it operates upon the steps of size-changes within the life cycle. Furthermore, it can operate on all the steps, or merely on some of them, for there are reasons why a step might be selectively neutral, and even a disadvantageous step could be retained if it were pleiotropically connected with a particular advantageous step. Also, modifier genes may emphasize some of the pleiotropic effects and minimize others. Thus it is evident that the whole pattern of steps (how they are grouped, how they mesh and interweave with one another) is a matter of prime concern in modern biology.

Populations differ from life cycles in that instead of fluctuating they tend to maintain an equilibrium. This is selectively advantageous, and there is some evidence that genetically determined social factors play a part in this homeostasis; therefore these population factors, along with all others connected with social cooperation, are introduced and eliminated in the life cycles of the individuals, even though their effects are extended to the whole population. For this reason, among others, the life cycle is considered the unit of evolutionary change.

Thus far we have been talking in very general terms about the kinds of changes in the steps which selection has produced. We have indicated that they are changes that occur during the life cycle beyond the minimum metabolic cell, and that some of these changes have effects which can even be seen in populations and societies. But it is possible to make a more systematic subdivision of these changes in the steps or variations; that will be the basis for the remainder of the chapter.

The system of subdivision is extremely simple. We can begin with the question of the amount of variation. This is an essential point because, as will be shown, without careful control of the amount of variation it would be impossible for evolution to occur. Therefore the amount of variation is itself a matter that is controlled by natural selection. We have repeatedly emphasized that certain properties, such as the moment of meiosis, are unimportant in the unfolding of the life cycle but may be exceedingly important in the matter of evolution. Here we will see specifically that it will affect the degree of variation.

The second type of adaptive change is in the kind of step. This really is the mainstay of orthodox evolutionary biology. There are qualitative differences between organisms, and the variety of these qualitative differences constitutes in fact the whole great panorama of animal and plant evolution. Here we will differ slightly from orthodoxy by stressing function (analogy) rather than phylogeny (homology) and by subdividing the kinds of steps on the basis of whether they belong to the period of size increase, size decrease, or size equilibrium in the life cycle.

The third and final category of adaptive change involves the number of steps rather than their nature; it is here a matter of quantity as opposed to quality. Since the number of steps is related to size, it is important to see what the adaptive consequences are.

In these three categories we have divided the kinds of changes that can occur into (1) the control of the frequency of change,

(2) the quality of the changes, and (3) the additive quantity of the changes.

THE AMOUNT OF VARIATION

The characteristics of an individual organism (throughout its life cycle) are determined by the particular combination of genes in its genome along with all the key cytoplasmic components. The character of a population is determined by the collective phenotypes of all the individuals. The individual and the collective phenotypes in turn are determined by the reservoir of genes in the whole population. But the relationship may not be a simple one, for the different genes which can be drawn from the reservoir in the population can exist in any one individual in a wide variety of combinations. Therefore there are as many possible phenotypes as there are possible permutations of gene combination in the genome, a vastly greater number than the number of genes in the population.

The result is that besides the production of mutation, great importance lies in the method by which the mutations are shuffled and recombined in the individuals. The array of phenotypic variation must neither be too small nor too great; it must be controlled by selection so that selection can effectively take place. This is an exceedingly important principle which many earlier authors refer to as the principle of compromise. It involves a control of the variation flexibility: too much variation will make it cumbersome and difficult to follow good leads for they will be obscured and clogged by the surfeit of new suggestions, and too few will make evolutionary progress unnecessarily slow.

It is our present task to see how this control of variation is achieved, and in particular how it is affected by the size of the organism and the nature of the life cycle. From past work we have a curious one-sided picture of variation in evolution. We have large amounts of information about genetic recombination, mutation frequencies, rates of change of the gene pool in a population, etc., but very little on the relation of the size and type of life cycle to its variation-producing mechanism. We assume that certain types of recombination systems have advantages and disadvantages in their shuffling for natural selection and that these advantages and disadvantages bear some relation to the structure of the organism and the structure of its life cycle. But exactly what the relations are is a matter that needs penetrating exploration. This subject is difficult to find discussed in any one place in the literature, although Stebbins (1950) is not only aware of the problem but helpfully analyzes a number of its aspects.

The plan here will be first to consider phenotypic variation as a means of providing flexibility. It will be seen that as an adaptation this has many advantages, but it is not a means of evolutionary change. Next we shall examine the various kinds of recombination (including its total absence). Since the important matter is the amount of recombination, each of these systems will be considered in terms of its flexibility and control. It is also possible to show that control is exerted by the generation time or the duration of the life cycle, as well as the type of life cycle. In fact, the number of elements affecting the amount of recombination and variation is so large that it is not surprising that different organisms with their particular life cycles have worked out different solutions to the problem.

Genetic and phenotypic variation

We have already seen that there are at least two kinds of variation: genetically fixed variation of the standard sort, and range variation in which the genes control the range of variations within a population but not the characteristics of any one individual. I should now like to introduce a third type (another kind of phenotypic variation) in which an individual can respond in a number of different ways depending upon the environmental circumstances and stimuli. To cite a very obvious example, certain plants such as *Myriophyllum* will have a different leaf form when grown underwater than when grown in air (review: Sinnott, 1960). It is even possible to obtain both leaf forms on half-submerged plants (Fig. 16). There are many other cases among animals as well as plants, as Schmalhausen (1949) has shown, and they all show the same flexible response to different environmental circumstances. In this case what is inherited is the ability to respond in different ways. We might call this kind of variation, to distinguish it from range variation, "multiple choice variation."

All three types of variation are gene-controlled. Two are rela-

FIGURE 16. An example of multiple choice variation. A species of *Myriophyllum* in which the submerged leaves have a totally different shape from the aerial ones. (After Fassett)

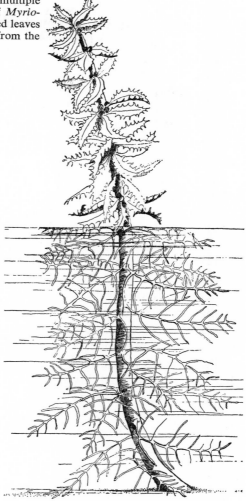

tively independent of the environment, one being rigidly controlled with the identical end result or phenotype, and the other being controlled within a range so that the nature of any one offspring is not predictable except within that range. The third is rigidly controlled along two or more alternative pathways, and the decision as to which path is followed is entirely determined by environmental stimuli.

Flexibility in changing environments can be achieved two ways, either by selection of many fixed variants or by possessing multiple choice variants. The result in each case will be quite different. In the former the genetic character of the population will change and evolution therefore will be occurring. In the latter, no matter how the environment changes, the individuals (within limits) will cope, and this they can do without any genetic changes whatsoever. Therefore we must conclude that multiple choice variation is not in itself a method of evolutionary change; it is an end-point in evolution, in fact an adaptation to changing environments.

A very interesting example of these two ways of producing flexibility in changing environments is found among bacteria and discussed recently by Pardee (1961). He begins with the assumption that rapid growth rather than metabolic efficiency is a significant adaptive advantage to the bacterium. This being the case, if all or most of the alternate metabolic pathways are stripped away, then in the presence of the proper substrate the growth rate will be especially high. However in any new environment the chances of such cells accommodating to a new substrate by the induction of the proper enzymes and pathways is most unlikely. On the other hand other bacteria keep a wide range of possible enzymes that can be increased readily by induction, and what they gain in flexibility they lose in speed of growth in any one circumstance. Both the flexible and rigid systems are gene-controlled and both exist in nature. Therefore it is advantageous to be restricted and streamlined as well as flexible and accommodating, and each has its disadvantages. Each case is adaptative, and either may change by mutation or recombination since they are both gene-controlled.

Another striking example of multiple choice variation may be seen in the antigens of the cilia surface or *Paramecium*. It has been shown by Sonneborn and Preer (review: Beale, 1954) that there are genes that govern all five possible antigens of the cilia, but which of the five appears (and only one is possible for any one organism) is dependent upon the external temperature at a critical moment of the life cycle. Once that antigen is established, the other four possibilities are automatically excluded. This is a multiple choice in which the multiple is five. While the mechanism is a perfect illustration of the principle, the adaptive significance of any particular antigen is hard to grasp.

What we have really shown here is that there are two levels of variation and variation control, one at the genetic level and one at the level of the phenotype. But the genetic level always comes first; it is the primary value. Without the appropriate gene the system of phenotypic variation cannot exist at all, and any change in the gene will change the method of phenotypic variation. However, changes in the phenotypic variation can have no effect on the gene. Therefore the control of variation is first and foremost a genetic problem.

Total absence of recombination

More and more unicellular organisms have been shown to possess some sort of recombination system, but nevertheless there are certainly some unicellular organisms that totally lack re-combination, as well as some multicellular ones. The latter case, which will be discussed presently, represents instances where a sexual ancestor gives rise to an asexually reproducing descendant for the apparent purpose of preventing any further variation and capitalizing on the great adaptive advantages of a particular genome in a particular environment. In nonsexual unicellular forms we presume (as does Boyden, 1954) that this is the primitive condition and the sex is secondarily derived.

The majority of the totally asexual forms are very small, haploid, and are characterized by short generations. Bacteria, which are the most important example, may produce many generations in a relatively brief interval of time, and since the mutation rate is geared to cell cycles, one can predict that after a certain number of generations a new mutant will appear. If the population is large enough and the turnover sufficiently rapid, the rate of appearance of new mutants will be considerable. All of those that are viable will produce progeny, and as a result the whole population, which may have started as a clone from a single cell, may be genetically very heterogeneous. It is obvious that if a sudden shift in environmental conditions occurs, some of the mutants might be successful in coping with it and others not; the result would be a surge of growth in some mutant types and not in others.

The only method of passing on genetic information in these forms is by direct replication of the genome. If a mistake, a mutation, appears it will without prejudice be passed to the daughter cells and perpetuated. There is no way of adding or subtracting mutant genes from the cells except by mutation and back mutation. A particular cell may have two or more mutant genes, but these can exist only in the combinations in which they originally appeared. No reshuffling is possible. Clearly then, as a variation system this has severe limitations and can exist only by virtue of the shortness of the generations and the magnitude of the turnover. The only real control of variation is the mutation rate, and there is no method of amplifying this for the phenotype by reshuffling. The fact that this system exists only in a few lowly and minute forms is perhaps a reflection of its fundamental inadequacy to provide sufficient basis for evolutionary change. This is supported by the fact that even bacteria have developed recombination systems.

Many higher sexual forms have close relatives that are asexual. The transition from sexual to asexual is a violent form of recombination suppression, for in one step all shuffling is eliminated. In this way evolutionary progress is brought to a grinding halt, but if a particular organism finds itself in effective equilibrium with a stable environment, the immediate advantage may be great. To understand this extreme form of control, two things must be remembered. Not all organisms are moving forward at the same rate; some are evolving rapidly and others may be to varying degrees static. If an organism has found an ideal, constant environmental niche, it is not surprising that sometimes sexual reproduction disappears and the successful recombinant is permanently preserved by asexual reproduction. This apparently is just what has happened in the buffalo grass of the prairies, which reproduces primarily by vegetative runners, and this is safe provided the environment remains reasonably the same and we do not enter upon a new ice age.

Another important point is that many organisms possess both a sexual and an asexual development in different parts of their cycles, which are geared to different environments at different seasons. Take, as an example, the case of the green alga *Volvox* (Plate 9) or *Hydrodictyon*. In the summer months these forms are likely to be in a lake or a pond that will provide relatively constant environment, at least for the season. If a particular variant has found success in this environment, it is disadvantageous for it to go into sexual reproduction, and it is far better to propagate as rapidly as possible asexually. Recombinants would be desirable only when the colonies are likely to be exposed to new environments. The very same situation applies to hydra, another fresh-water form (Plate 19). Asexual budding occurs in summer, but when the unfavorable conditions of the fall appear, gonads develop.

There is still another method of asexual reproduction, and that is by the modification of sexual reproduction so that no recombination is possible. Parthenogenesis or apomixis is the method, and it is common among both plants and animals. The classic instance, discovered by Charles Bonnet in the eighteenth century, is that of some aphids. During the summer months some species reproduce parthenogenetically; and in the fall the males take part, mating occurs, and the eggs that last over the winter are sexual, producing variable offspring to meet the uncertainties of the next warm season. A very similar situation is to be found among the water fleas (*Cladocera*).

Besides strict parthenogenesis, there are intermediate cases which lie somewhere between true asexual and sexual reproduction. For instance Cleveland's (1956) work describes the flagellates in the intestine of wood roaches, where at molting the protozoa are extruded in the faecal pellets and at the same time the eggs, which have been laid previously, begin to hatch and take on their intestinal fauna. Following the argument of Sonneborn (1957), we might presume that originally this discharge was accompanied by full blown meiosis, which Cleveland has shown is stimulated by the molting hormone ecdysone. This is an advantageous moment to produce new variations since new hosts, new environments, will be entered. However, the meiotic process in present-day roach flagellates is, according to Cleveland, so modified that effective recombination is reduced. From this Sonneborn suggests that the ideal situation with normal meiosis was ancestral when rapid evolutionary changes were in progress, and that we have today a degeneration to reduce recombination so that the successful roach-flagellate symbiosis can be maintained in stability. From the point of view of evolution, this ossification of meiosis acts in the same way as parthenogenesis and asexual reproduction.

If we return to the general relation of asexual reproduction to size and the whole life cycle, in unicellular forms we invariably find one or many cell divisions or mitotic cycles between the rarer sexual cycles. In multicellular organisms, while the mitotic cycle continues for the individual cells, the whole organism may or may not indulge in asexual reproduction. While we understand the adaptive advantage of asexual reproduction, we cannot properly correlate its presence or absence with any particular kind of organism or any particular kind of habitat. The possible exception to this is that vertebrates do not possess it at all, while it is present in all major invertebrate groups, tunicates, and throughout the plant kingdom. But it is also true that in all groups of organisms there are many forms which lack sexual reproduction. Therefore sexuality and asexuality seem to go side by side, except in the case of the vertebrates where only one is present. Perhaps the best way of looking at the matter is to think of the evolutionary significance of asexuality as a method of reducing variation. One might expect a total disappearance of asexuality in modern groups where the rate of evolution has been very rapid, as in the vertebrates, while in the groups which have moved at a steadier and more sedate pace the presence of asexual reproduction is fitting and expected. Such a correlation, however, is unsatisfactorily vague.

Another possibility suggested by Stebbins (1950) is that the absence of asexual reproduction is directly correlated with complexity. Vertebrates are indeed the most complex of all organisms, and he suggests that the more complex the form, the more difficult it is to find adaptive recombinations because there are so many complex developmental permutations that have to be satisfied. For this reason a large amount of recombination is constantly needed if there is to be any evolutionary change at all. Whether this riddle is to be answered in terms of complexity or rate of evolutionary change or both is not clear, but both these elements are undoubtedly important and affect the control of variations.

We have begun by considering instances where there is a total lack of recombination. Next we can look at a number of different systems where recombination is achieved in various ways with varying amounts of success and efficiency.

Cell recombination

As has been discussed elsewhere in some detail (Bonner, 1958; Filosa, 1962), cell recombination can occur only in aggregative organisms. Since in myxobacteria or cellular slime molds cells of different genetic constitution may come together by aggregation, the newly formed cell mass has been termed a "heterocyton." In fact it has been demonstrated by Filosa that, at least under laboratory conditions involving serial transfer by mass spore inoculation, more than one cell type will be carried in the cell population.

If, upon fruiting, each one of these cells in the form of a spore is scattered separately to form a clone, there is obviously no possibility of recombination. Recombination requires passage from one mixture of genomes to another, and therefore genetically different cells must enter an aggregate (e.g. A+B+C+D+E) so that upon dissemination of the spores they came off in groups of different combinations (e.g. A+B, B+C+D, A+D, D+E, etc.). Certain of these recombinations may have particular advantages in particular environments and thus be maintained over a number of life cycles, as Filosa has shown for one particular variety of cellular slime mold.

The situation is the same for the coccus forms of myxobacteria, but those species that form cysts, such as *Chondromyces*, have a further advantage (Plate 5); many hundreds of cells, which can be genetically diverse, are grouped into a cyst which may thus contain a recombined set of genomes. As a result the progeny of one cyst will not be a genetically pure clone but a mixture, and the mixture may be different for each cyst.

Heterocytosis, as one can plainly see, provides the crudest sort of recombination. It is interesting that when two cell types are mixed, the phenotypic expression may be of one type only; a cell type can express dominance in a cell mass (Filosa, 1962). Particular combinations of cells may have particular advantages, but these can hardly be compared to combinations of different genes in one nucleus. It should be added that ordinarily in the cellular slime molds (and undoubtedly in the myxobacteria) the nuclei are haploid. However, Ross (1960) has shown that diploid cells do appear occasionally, and this raises the possibility that at least in these cases there are further possibilities of recombination. Sussman and Sussman's (1963) recent demonstration that some strains can readily shift from the haploid to the diploid form and back reinforces the idea that these slime molds also occasionally have some sort of true sexual recombination, but unfortunately thus far the evidence is insufficient. If one compares any sexual system, which has different gene combinations, with heterocytosis, the difference is largely quantitative, for in heterocytosis the number of possible combinations is severely limited as compared to those within a single cell in a true sexual system.

Nucleus recombination

Ever since the pioneer work of Hansen and Smith (1932; Hansen, 1938) it has been well recognized that imperfect fungi have a system of asexual recombination. These forms are capable of heterokaryosis, that is, by the fusion of genetically diverse hyphae there can be a mixture of nuclei. As with heterocytosis, it is thus possible to have more than one parent, and furthermore in the spore dispersal the conidia (i.e. macroconidia) have two or more nuclei, and therefore the new combinations are repackaged, with the result that each separate spore has different genomes to cope with a new environment. Also it is well known that these different nuclei may live together in balance over a long period, or in other cases one or a few of the nuclear types may be selected out. Furthermore, dominance is expressed in a heterokaryon, and the phenotype of a mycelium may be entirely that of one of the nuclear types and not the other. In all these respects they closely parallel the situation described for heterocytosis in the cellular slime molds.

Also occasionally, as Pontecorvo (1958) showed for *Penicillium*, the nuclei became diploid, but in this case he clearly established mitotic reduction back to haploid and was able to prove chromosome exchange and recombination. For many of these fungi, if not the majority, besides this parasexual system and the asexual system previously described, there is a true sexuality as well. It is only in some completely imperfect forms that recombination is achieved solely by the shuffling of whole nuclei into spores, and in these instances they are by comparison greatly reduced in the total number of possible genetic permutations.

Chromosome recombination

By far the most successful means of shuffling variation is that of chromosome recombination. This is successful in the sense that it is the most effective way of producing new combinations, and also in the sense that it is common to all groups of animals and plants.

There are, of course, two aspects to the shuffling. One is exchange of whole chromosomes which behave in a Mendelian fashion. The other is the actual exchange or rearrangement of parts of chromosomes as a result of crossing over. The former is relatively limited in its extent. The number of combinations is determined by chance and the number of separate chromosomes or linkage groups. The latter, since breakage can occur anywhere on any one chromosome, produces far more possible combinations, so many in fact that there are special mechanisms developed to suppress or control the amount of crossing over.

The whole matter of chromosome behavior in its genetic and cytological detail is in itself a vast subject. Furthermore some authors, notably Darlington (1958), and White (1945) (see Stebbins, 1950, for an excellent review), have examined the problem specifically in terms of adaptiveness and evolutionary change, as is implied in Darlington's phrase, "Evolution of genetic systems."

Since the standard sexual mechanism requires meiosis and fertilization for both the processes of chromosome shuffling and crossing over, it is clear that there must at some point in the life cycle be a diploid set of chromosomes and at another point a haploid set. We have been using size as a yardstick in life cycles, and we can now show another cause for size variation in that sexuality requires that one stage be at least twice the size of another. It is true that during most life cycles the size variation is far greater than this, and therefore this only applies to unicellular organisms. But even here normal cell division reduces the cell in half; the point really is that in meiosis and fertilization not just the dry weight is halved (or doubled) but the total amount of DNA, which is in a sense a more meaningful size change. It should be added that in heterocytosis there is also a size change associated with the recombination, for the shift from the aggregation to the dispersed spores (and the subsequent re-aggregation) involves massive size changes and reshuffling. In heterokaryosis there is likewise an increase in size by association while the heterokaryon exists, and a reduction when the conidia are formed and dispersed.

Thus far we have mentioned only those cases that involve recombination of whole chromosomes or parts of chromosomes in some kind of reciprocal exchange. All these fit under the heading of true sexuality, as do a number of cases where by accident the exchange may not be precisely reciprocal. In some cases during crossing over there is a loss of a piece of chromosome (deletion) or the addition of a fragment (translocation), but these are accidents in a normal reciprocal change.

In recent years Lederberg, Tatum, and Zinder have discovered a number of interesting cases of recombination in bacteria, none of which appear to be reciprocal. In transduction small pieces of chromosome from one bacterium, with the help of a bacteriophage, enters another and makes its contribution in its appropriate position on the parent chromosome. This may involve such small sections of chromosomes that it can hardly be considered chromosome recombination, but recombination of small pieces of chromosome. Even smaller units of change are involved in the classic transformation of Avery and McCarty, for here one or a few genes are altered by the external application of the DNA of a variant strain. This would indeed be change on the molecular level if one had any evidence that this kind of change occurs in nature; thus far it has only been observed by using artificial laboratory procedures. But even if it were shown to be a normal process, it would not be a method of recombination but rather a method of specific alteration, a kind of controlled rather than a random mutation. Transduction, however, normally occurs and produces recombined offspring, even though the exchange between partners may be completely one-sided.

Organism recombination

I have stressed that variation, which is all important for natural selection, is produced by mutation and is reshuffled in a variety of ways. It may be by the mixing of whole cells, of whole nuclei, of whole chromosomes, or of parts of chromosomes. As one proceeds along this scale the unit of recombination becomes progressively smaller and the possible permutations become very much greater. It is perhaps not surprising that also as one proceeds along this scale the number of organisms that possess the method increase strikingly. This is really an understatement, for all groups of organisms have a sexual system with chromosome recombinations, and in those cases where there is cellular or nuclear recombination there is usually (and possibly always) chromosome recombination as well.

Cellular recombination occurs only in aggregative organisms, but they suggest the interesting point that all truly sexual organisms, both unicellular and multicellular, have a kind of organism recombination that is of considerable significance. The recombination found in any offspring is largely the result of the chromosomal behavior of the male and female gametes that make it up, and also it is self-evident that it is determined by the particular genetic constitutions of the two parents. If one examines a population of individuals, each undoubtedly has a different genetic constitution and therefore it makes a difference which combination of parents comes together; for each pair would produce different offspring even without chromosome recombination. Therefore sexual organisms always recombine twice: once on the organism level and once on the chromosome level.

The result is that the size of the breeding population of any organism directly affects the degree of variability among the individuals; small populations are more likely to be inbred and genetically relatively homogeneous compared to large ones. The situation is in fact far more involved and interesting than this, as Sewell Wright has shown (see Dobzhansky, 1941), for with the reduction of the size of the population there may be a reduction in the total amount of variability due to random fluctuations in the frequencies of the genes. This will result in "random fixation" or "drift" of certain genes in a population. We may conclude, therefore, that one of the ways of controlling variation is by controlling the size of the population. In other words we again find that size is of critical importance, and we may conclude that recombination operates and is controlled at all size levels: populations, individuals, cells, nuclei, chromosomes, and parts of chromosomes.

Unisexuality versus bisexuality

Another major factor in controlling the amount of variation in the sexual phase of the life cycle is whether an individual of a species is bisexual or whether there are separate male and female organisms. Since both possibilities occur in different groups of animals and plants, there are a number of terms that have been elaborated to describe the same thing. For the one condition we have "bisexuality," "hermaphroditism," "dioecy," and "heterothallism"; and for the other we have "unisexuality," "monoecy," and "homothallism." For the sake of consistency and simplicity we will use the terms uni- and bisexual for all types of organisms in the following discussion.

There is some reason to believe that the adaptiveness of unisexuality is connected with the selective advantage of preventing inbreeding. Excessive inbreeding has the same result as asexuality; it produces homozygous individuals, that is individuals which are genetically identical. If an individual is of one sex only and has to find a mate, this automatically reduces the chance of a brother-sister mating and favors cross-breeding. We know that unisexuality alone greatly reduces inbreeding even though there is little evidence among animals, especially vertebrates, of any specific mechanism preventing brother-sister matings. Unfortunately concrete evidence on this score is rather meager except in the curious case of human beings where all societies (except for a few royal dynasties), no matter how advanced or how primitive, have an incest taboo. Yet so far as we can determine there is no physiological basis for this aversion; it is, we presume, a culturally acquired custom which happens to be biologically sound. So far as we know (but really we know very little), there are no social or biological barriers to inbreeding among other vertebrates, where it tends to be prevented by one of two phenomena: unisexuality accompanied by the movement of individuals, or bisexuality in sessile organisms which possess incompatibility factors.

Movement is necessary so that brothers and sisters or the parents and the offspring separate. Perhaps if any argument could be made for an incest-prevention mechanism among animals, it would be that there is an inborn instinct for families to break up after the period of childhood care. It is not known if there is such an instinct, but if there were it would favor outbreeding, provided of course that the organisms can move in order to separate. One could argue that there is a fair correlation between unisexuality and locomotion, the exception being largely among invertebrate groups such as some molluscs and some worms which are motile, and some unisexual plants which are sessile. We could also reasonably propose that the ability to move came first, and unisexuality evolved next as a simple adaptation to prevent inbreeding.

If we turn to bisexual organisms it is certainly true that they have gone to extraordinary lengths to prevent self-fertilization. The subject has been especially closely studied in plants and has been excellently reviewed in a number of places (e.g. Darlington and Mather, 1950; J. Raper, 1954). The mechanism in angiosperms may be classified into two major categories: those plants in which the male and female parts develop and ripen at different times and therefore cannot self-fertilize, and those in which there are genetically controlled incompatibility factors which prevent the pollen of a plant from penetrating into its own ovary. Both of these methods are gene-controlled and differ in different plants, indicating that the mechanisms are convergent and arose independently in evolution. Further work has been done to elucidate the system in Basidiomycetes, which is of quite a different nature. Both by doubling the number of loci concerned with mating and by greatly increasing the number of alleles at one locus, the statistical chance of brother-sister mating has been reduced. Incompatibility factors are also known to exist in animals; for example T. H. Morgan (1938 *et seq.*) showed that bisexual ascidians have genetically controlled means of preventing self-fertilization.

To return to our correlation between motility and unisexuality, we find support in the observation that nonmotile bisexual organisms such as plants and ascidians have elaborate systems to prevent inbreeding. Excessive inbreeding is clearly inadaptive, and by incompatibility mechanisms or by a system of locomotion the proper insurance against inbreeding is automatically maintained.

It is puzzling that in many groups of organisms, as in virtually all plants and many groups of invertebrates, there will exist both uni- and bisexual forms side by side. For example, there are species of *Volvox* which are unisexual and others which are bisexual, and the same could be said of *Mucor, Hydra,* and a host of other forms. In some cases both the uni- and bisexual species live in close proximity one to another. It may be, of course, that we are witnessing here the beginning of unisexuality as an incompatibility mechanism. It is also possible that while unisexuality has advantages as far as incompatibility is concerned, it has a selective disadvantage in that the likelihood that the opposite sexes can find one another at the appropriate time is automatically reduced. In the examples cited there are no elaborate devices for orienting one sex to the other, as among insects and vertebrates. Although the advantages of unisexuality are balanced by the disadvantage that the sexes may never meet, there is one further bit of insurance against catastrophe: all these organisms have asexual reproduction also, so that failure to find a partner may not be completely fatal to the future propagation of an individual in an isolated environment.

Some interesting and pertinent observations have been made by Sonneborn (1957) on different strains of *Paramecium aurelia.* There is apparently an excellent correlation between the degree of outbreeding and both the length of clonal life and the life span of individuals. The inbreeding strains are short-lived and the individuals come to maturity very quickly, while the outbreeders have an extended juvenile stage (during which they are incapable of conjugation) and an extended mature period. Sonneborn argues that the longer the life span and the period of immaturity, the more likely it is that the individuals will go some distance and find an individual of the opposite mating type. To further insure that this occurs, there are numerous mating types in the outbreeding strains as compared to those in the inbreeding strains. Presumably the short-lived inbreeders remain relatively fixed genetically, while the long-lived outbreeders have far greater opportunities for recombination and therefore evolutionary change. However it must be remembered that over a long span of time there will be many more meiotic divisions and conjugations among the inbreeders than among the outbreeders, and this may to some extent compensate. In any event, both systems must be adaptive for they both exist.

Haploidy versus diploidy

It is striking that the ploidy of organisms varies considerably, and it is our aim to show here that this has significance as far as the degree of recombination is concerned. There are a number of variables, the most important of which is the moment when fertilization and meiosis occur within the cycle. Some of the different permutations for various organisms are shown in Figure 17, and these will be discussed in some detail.

First we may compare those forms in which fertilization immediately follows meiosis and those in which fertilization immediately precedes meiosis. In the former the haploid state is very brief, extending only to the life of the gametes, while in the whole remainder of the cycle the organism is diploid. In the latter case it is the period of diploidy that is short, and the bulk of the life cycle is haploid.

Organisms which have a very short haploid and an extended diploid phase may conveniently be called diploid. This is apparently a very successful type of cycle, for all the major groups of animals and the gymnosperms and angiosperms in plants have this system. Since it is found among what are generally regarded as the most successful, most elaborate, and most advanced forms, it is presumed to be the most suited for evolutionary progress. We would like to know the reason why.

The accepted answer comes from the population geneticist. By remaining diploid throughout most of its cycle, a heterozygous animal or plant can harbor a far greater number of genes than can a haploid organism. This means that in a population of heterozygotes the total number of different genes in the gene pool is greatly extended, and therefore the possibilities of finding new combinations to meet new exigencies in environmental circumstances are great. In other words the extent of recombination is

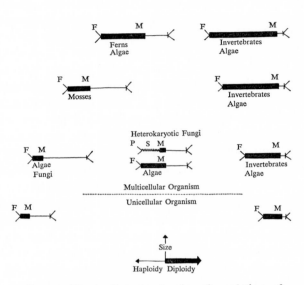

FIGURE 17. A diagram showing the relation of haploidy, diploidy, and size. The size of the organism is indicated by the length of the lines (which represent life cycles). The thick line is the diplophase, the thin line the haplophase. The length of the line is proportional to the generation time or duration of the cycle. F = fertilization, M = meiosis, P = plasmogamy, S = karyogamy.

increased by diploidy. The difficulty is that besides the useful and beneficial genes, deleterious and lethal genes will also be harbored in diploids; but if the reproduction rate is sufficiently high, the periodic expression of any undesirable gene may not be of any consequence to the population as a whole.

The argument is based purely upon circumstantial evidence, fanned by the feeling of great need for an explanation of the success of this type of cycle. Yet despite the fact that it has not had any kind of experimental proof, it seems to be reasonable. Sometimes objections are raised on the basis that the gametes are haploid and that during this period the unmasked genes should have deleterious effects. But this is no objection, for only lethals which act during the life of the gametes can appear at this stage; those that damage some developmental stage can appear only during the diploid phase and then they will be masked by their allele. Another objection is the case of wasps and bees where the sex difference is based upon whether the eggs are fertilized or not. The males are produced parthenogenetically and are therefore haploid; yet this does not necessarily result in a high mortality in males. One would be forced to reply to this argument that these Hymenoptera do not obtain the same benefits from diploids as other organisms. If the males are haploid the total number of the genes in the gene pool must be reduced. Perhaps as a result evolution in the Hymenoptera has been slower, a contention that would be hard to prove or disprove.

Besides large and elaborate diploid organisms there are also some minute ones, for example many sexual protozoa (e.g. ciliates) and diatoms. Therefore there is not a perfect two-way correspondence between size and diploidy, although the exceptions, as Stebbins (1950) points out, are particularly complex for small forms.

In the case of haploid organisms, those in which fertilization is immediately followed by meiosis, it is harder to find a rationale for the adaptiveness of such a cycle. There are unicellular forms which follow this pattern (e.g. *Chlamydomonas*, Plate 3), and it is found in a curious modified form in fungi, although the result with fungi is that even though they are in fact haploid, they are functionally diploid or even polyploid. This statement needs further elucidation.

In Phycomycetes such as *Mucor* or *Rhizopus* there is a fusion of two individuals (in heterothallic forms) to form a zygote or zygospore. The gametes, which consist of no more than sealed hyphal tips, are multinucleate, and the zygospore has more than one zygote nucleus. Meiosis ensues upon germination, and the new mycelium that issues forth is made up of a population of recombined nuclei. Thus each individual *Mucor* houses a population of haploid nuclei, of which there could be and undoubtedly are more than one genetic type. Therefore it does not have just two sets of genes like an ordinary diploid but more, although these all lie in haploid nuclei. The evidence that they are functionally diploids, in the sense that one gene from one nucleus can dominate the effects of its heterozygous allele in another haploid nucleus, comes from work in the Ascomycetes, where this kind of genetically diverse population of haploid nuclei is even better understood.

This is largely due to the detailed work on *Neurospora*. In the Ascomycetes fertilization involves two distinct steps: one is the fusion of the hyphae and a mixture of the nuclei of the parents (plasmogamy), and the second is the actual fusion of pairs of haploid nuclei (karyogamy). In this instance fertilization itself has undergone heterochrony, and as a result of plasmogamy there is an extended state of mixed nuclei of two (or more) parents called heterokaryosis. If one parent is able to synthesize a particular substance and the other is not, the heterokaryon of these two, which has some haploid nuclei from both, can be tested with regard to its synthetic abilities. As already mentioned, it is found that simple Mendelian dominance can be expressed here as readily as in the diploids. There is also the interesting question, which lies beyond the scope of the present discussion, of what ratios in the number of the two nuclei are needed for the expression of dominance, and one sees clearly that there are methods of maintenance of these ratios. There are also interesting problems connected with the movement of the nuclei from one part of a mycelium to another following plasmogamy.

The period of heterokaryosis may extend over long periods of growth and mycelial enlargement; it may represent a major portion of the life cycle. This polyglot haploid condition is terminated only by the final paring of the genetically different nuclei in the ascogenous hyphae, and their fusion. Meiosis follows immediately, usually with the production of eight ascospores.

The only exception to this general pattern is the interesting parasexuality discovered by Pontecorvo (1958). Here, by some quirk of division, the vegetative nuclei become diploid, and these diploid nuclei periodically undergo a recombination and a return to haploidy by mitotic reduction. It is an extra system of recombination superimposed upon the normal one. This type of sexuality can only serve in a multinucleate organism such as a fungus, where any nucleus has a good chance of becoming part of the germ line. In the case of higher uninucleate forms, such a recombination system would be lost or would find itself in a dead end if it produced any local changes in the soma.

The Basidiomycetes, which presumably are derived from the Ascomycetes, follow much the same system. Here the haploid, primary mycelia fuse in plasmogamy, and the resulting dikaryon has one novelty that is absent in the Ascomycetes: the nuclei from the two parents come together and lie side by side in close association. It is as though there were a need for further insurance that the essentially diploid character of the haploid mycelium be maintained, at least functionally. The dikaryon period again may be greatly extended in time, leading finally to syngamy and basidospore production, a process closely resembling ascospore formation.

All the changes described on the sexual part of the life cycle of these various fungi are presumably related to the amount of recombination. The systems and their variations are such that the adaptive degree of recombination occurs. It should be added that along with the sexual and the parasexual recombinations there is yet another. As mentioned previously, the fact that a hypha can have the nuclei of two or more parents, and that the asexual conidia can collect there in different combinations (as originally recognized by Hansen and Smith, 1932), means that heterokaryosis alone provides a system of recombination. There certainly does not seem to be a dearth of mechanisms of variation production among fungi.

Also, as has been shown, in the cellular slime molds there is a parallel nonsexual system of recombination. Again some of the aspects of diploidy are imitated, although obviously the efficiency of such a system in handling variation for natural selection is negligible compared to a truly sexual system.

Returning to the question of haploid versus diploid organisms, we have seen in the case of fungi that even though they are literally haploid throughout most of their cycles, they are functionally diploid. Does this mean that there are no truly large and complex haploid sexual organisms other than the males of Hymenoptera?

A possible exception might be the alternation of generations found in the algae, the Bryophytes, and to a lesser degree the Pteridophytes. Here it is possible to have a haploid generation followed by a diploid one. Let us examine the sea lettuce, *Ulva*, which is a reasonably large form and has a well-developed alternation of generations (Plate 10). In this case it is impossible to distinguish between the haploid and the diploid thallus except by a close examination of cell or nuclear size or by actual chromosome counts. The inference is that if a young plant is haploid or diploid, its development is affected in no obvious way. The diploidy in this case cannot be of any advantage in the enlargement of the gene pool; its genetic hand has to be held out in the open every haploid generation.

In many algae, such as the more complex browns, the diploid stage becomes increasingly prominent, until, as in *Fucus*, the haploid stage disappears completely (Plate 12). There is a beautiful evolutionary trend from the Bryophytes up through the Pteridophytes showing a diminution of the gametophyte and an elaboration of the sporophyte (Plates 15, 16). In the case of mosses, for example a haircap moss, the bulk of the plant is the gametophyte. It starts as an algal-like protonema which buds into the erect moss, at the tip of which antheridia and archegonia are formed. From fertilization a diploid sporophyte (the hair cap) grows up from the gametophyte and, following meiosis, produces spores, each one of which can start a new protonema (Plate 15). It has been possible to induce the parthenogenetic initiation of the sporophyte, making it haploid as a consequence, yet it is nevertheless perfectly formed. In other words the previously mentioned advantages of diploidy do not seem to be exploited in any way.

In the ferns the large part we see is diploid and, of course, spore bearing, while the gametophyte which results from the germination of the haploid spores is an inconspicuous liverwort-like prothallus (Plate 16). In the horsetails and the clubmosses the gametophyte is smaller, and finally in the angiosperms and gymnosperm (Plates 17, 18) it is so reduced that it would be unrecognizable without a knowledge of the ancestry of the plants, for it is so wedded to the sporophyte tissue.

Therefore again in plants we return to the notion that diploidy is correlated with size and complexity; it appears to be a reasonably valid and consistent correlation. It might be added that among higher plants of many species there has also been a tendency to produce polyploid forms. The advantages gained by this further duplication, however, are quite a different matter from the original advantages of diploidy. Polyploidy serves the same role as mutations; it is simply a fresh source of variation. Although it is true that polyploids can hold more genes and increase the gene pool in a population, this attribute probably plays no role in the selective advantages of polyploids. This advantage seems to be more immediate, just the way the advantage or disadvantage of a particular mutant may be selected for or against.

Conclusion

We began with the idea that the extent of variation is of vital importance in the process of evolution. Too little or too much variation impedes progress: there is an optimum that is in itself controlled by selection.

There are many different ways of affecting the amount of variation, and these vary with the nature of the life cycle and the size and complexity of the organism at its point of maximum size. It is possible to list some of the key factors which influence and control the amount of variation:

> The mutation rate
> Cell shuffling in aggregative organisms and nuclear shuffling in heterokaryotic fungi
> The number of chromosomes
> The amount of crossing over
> The size of the population
> The degree of inbreeding *vs.* outbreeding and all the mechanisms that affect this
> Haploidy *vs.* diploidy

The important question is how these and possibly other factors affecting variation control are correlated with the characteristics of particular life cycles. For example, if the generations are very short, as in microorganisms, it is not surprising to find that often the sexual events are few and far between, for mutation here takes on a far more important role and its results can be largely handled through rapid asexual cell divisions.

The converse is also true; with increased size and complexity organisms tend to rely more heavily on the sexual system (which can be controlled with far greater precision). Furthermore, with the size and complexity trend there is a trend towards diploidy which increases the reservoir of variant genes.

Another aspect of the life cycle that can be correlated with the mechanism of variation control is the degree of motility. Effective movement of individuals favors outbreeding and therefore increased variability. In general, motile organisms tend to be unisexual, and sessile organisms have devised elaborate incompatibility mechanisms to prevent inbreeding.

Finally, for any one organism it may be selectively advantageous to be variable at some moments, or seasons, and fixed at others, and in these cases there may be a series of asexual cycles with a periodic sexual one at a key moment when variation is required.

All these factors, and others we have undoubtedly failed to mention, produce a complex picture. There are many ways to achieve and control variation, and for any one organism a number of these are usually involved simultaneously. This is not the kind of problem that fits within the conventional and relatively simple postulates of the population geneticist, and we are indebted to Lewontin (1961) for a recent first step towards a mathematical attack on this strategic problem. As he so correctly appreciates, it is the kind of problem that lies within the province of the theory of games. We want to know how the concert of all the variables is coordinated and conducted so that we can predict the optimum amount of variation and the optimum method of producing it for any life cycle of any particular organism. Furthermore, we must expect, as is so often the case with living organisms, that a particular problem of strategy or compromise has more than one possible effective solution, and that this in part explains the diversity (and convergence) we find in the living world.

THE KIND OF STEP

The period of size increase

We are concerned here with what might be called the traditional content of evolutionary biology, the diversity of organisms resulting from evolution. Most biologists assume that life started on earth from one or relatively few beginnings, and that once the cell was established, all the fantastic diversity, all the millions of different species that have existed in the history of the earth, were produced by variation and natural selection. The process of altering steps in a population by making innovations and eliminations in life cycles has been extraordinarily successful in producing new kinds of steps and sequences of steps.

There has been some discussion whether the innovations that produce these new steps can possibly be utterly random, or whether there are some limits or some direction. It has been pointed out by Blum (1962) that there are certain physiochemical restrictions that are imposed upon the mutation process. Waddington (1957) has suggested that while mutation is in no strict sense directed, the developmental processes are so constructed through the formation of canals or rigid pathways that the pleiotropic effects of any new mutation can be influenced and to some extent controlled. Both these views, as are those of the vast majority of biologists today, are strictly orthodox in the sense that there is no postulating of any directive force which governs the kind of mutation that occurs, such as was popular with some scientists during the early part of this century to explain orthogenetic trends in some evolutionary lines. This question is at the moment beyond proof or disproof, but for simplicity we will assume that within the limitations imposed by physio-chemistry and the developmental machinery of the organism, mutations are not in any sense directed or predetermined.

This means that all innovations since the first cells, from angiosperms to liver flukes, from whales to lichens, are derived from the ability to make new steps and to add one step to another so that chains of steps can be extended.

If we set ourselves the task of classifying the new steps into some universal system of categories, there are so many different kinds of steps that the task seems almost impossible.

It could be done by describing and classifying the final result of the period of size increase, that is, the adult. This does not classify the steps but merely the results of the steps. This practical solution of an impossible problem has been the most satisfactory one devised, even though the problem of describing the limits of any one species is by no means always easy or obvious. Nevertheless, from many points of view, this has been an important tool in biology, and the systematists are continuing to make contributions.

We shall now look at a scheme of classification that utterly lacks the usefulness of the method of modern taxonomy but has, to make up for this loss, a rational simplicity that may be helpful in gaining a deeper understanding of the structure of biology.

In the first place, the steps are subdivided into the periods of size increase, size decrease, and size equilibrium. Within the period of size increase, which is after all the period where the vast majority of steps occur, we can further classify the steps on the basis of function or division of labor. Steps will lead to the formation of a particular structure, and that structure will then perform certain functions within the life cycle. Major categories of living functions would be locomotion, feeding (energy conversion), coordination, and reproduction. For each of these, in any one life cycle, there are certain specific structures which, by division of labor, are associated with the function: muscles and limbs for locomotion, alimentary canals and all their accessories for feeding, and so forth.

For any one function there will be many kinds of steps; this is a very large and crude category of steps. For instance, locomotion may occur in vertebrates by running, swimming, or flying, and for each method there are contractile muscles, bones, and tendons. Therefore, within the general category of locomotion there are subdivisions characterizing more specifically the particular variety of the function. In this case there are structural differences between a wing and a leg, while there are common features in the muscle, bone, and tendons. If we look within the category of swimming, we find not only vertebrates but protozoa as well. Here, while the latter do not have muscle, they have similar contractile protein which acts within the cilia. Again, there are things in common and things which differ if we compare all swimmers. One assumes that in evolution one began with a property at the cell level—in this case contractile protein—which became emphasized and exaggerated by division of labor or differentiation. During the course of this elaboration the steps developed into alternate pathways.

In some cases, which we call convergent, the same function is achieved by somewhat different structures: the wing of a bird or a bat or an insect. This shows that a single function can be achieved by a number of different structures, and therefore a functional classification of steps is imperfect in this sense. But this imperfection is itself important because the evolutionary opportunity was for the function and the fact that there are a number of different kinds of steps which can fill the same function is of great interest.

The kind of step is then to some extent governed by the function, and it is also, as was just pointed out, partly governed by the division of labor or differentiation. The more extensive the division of labor, the more elaborate the steps, and therefore the greater the number of new kinds of steps. This means that we cannot, except in abstraction, really separate the question of the number of the steps from the kind of steps, for we see that if there are more steps there are also more *kinds* of steps through differentiation. Since size is closely related to the number of steps and varies with complexity, we must always expect that quality and quantity go together, or at least one cannot have an increase in size (number of steps) without having a corresponding increase in complexity (kind of steps), as can be seen from the crude graph shown in Figure 2.

This is not the place to try to make a complete chart of all the functions and all the corresponding kinds of steps, even if this were possible. The important point is to show that evolutionary opportunity favors certain functions, and this has the effect by mutation and natural selection of permitting new steps that appear and establish themselves. These steps are connected with more efficient division of labor, which one presumes is adaptive. The period of size increase, which is the period of the vast majority of steps, appears to be centered around the notion of functional-structural stability by differentiation.

The period of size decrease

As we have seen, size decrease can be roughly subdivided into abrupt and gradual categories. Abrupt size decrease is more than an adaptive advantage, for in some ways it is totally unavoidable. The most obvious reason is that to make a cycle both growth and separation are required, and separation could hardly occur without a size decrease. Thus to some extent both size decrease and size increase are simply integral parts of the life cycle and they are adaptive at least in the sense that the whole cycle is adaptive. But in populations, where elaborate arrangements are made to keep the size constant, size increase and decrease are either absent or appear by chance; at least they are not an integral aspect of the population.

There is an additional related reason that size decrease is inevitable in many life cycles (Bonner, 1958); in all sexual forms, which constitute the large majority of animals and plants, the system of recombination requires that single male and single female cells undergo meiosis, and fertilization must involve one haploid gamete from each. Therefore, no matter what the maximum size during the life cycle, there must be a point at which the organism consists of one cell. This means that the larger the organism, the longer the period of size increase, and the more dramatic the period of abrupt size decrease.

Not only is this true of all sexual forms but of many asexual ones as well. Spores are also minute, although their small size is associated with dispersal rather than with the demands of meiosis and fertilization. The only instances where the reduction is less dramatic are those cases of asexual reproduction where an organism divides in two or gives off a bud. In single cells the product of division is still, of course, very small, while in multicellular forms such as hydra and various flatworms the point of minimum size after division or budding may be larger (Plates 19, 20).

As will be remembered, the principal method of gradual size decrease is senescence. Ever since the days of Weismann (1891) there has been a considerable debate on the question whether the process of senescence is adaptive. Here we shall simply assume that it is and examine the various possible qualities or traits of senescence that could be of adaptive significance.

Weismann began the discussion with the suggestion that senescence and death are adaptive because worn out individuals are valueless and even harmful to the species. As Medawar (1957) and Comfort (1956) have pointed out, this argument is circular; perhaps it could more effectively be stated that a particular life span or mean duration of life is adaptively advantageous for any one species, and in some cases this is achieved by senescence. Here we will consider three possibilities as to how the duration of the life span could be adaptive.

The first (and probably the most significant) is related to a concept of R. A. Fisher (1958, review: Slobodkin, 1961). It is the notion of the *reproductive value* of the individual, which is defined as the diminution of increase in a future population by removing a single individual of a specific age from the population. In man (which is probably representative) the reproduction value rises to a peak shortly before twenty years of age and then falls off rapidly afterwards (Fig. 18). If this curve is considered from the point of view of selection, it is clear that the higher the curve the greater the opportunities for effective selec-

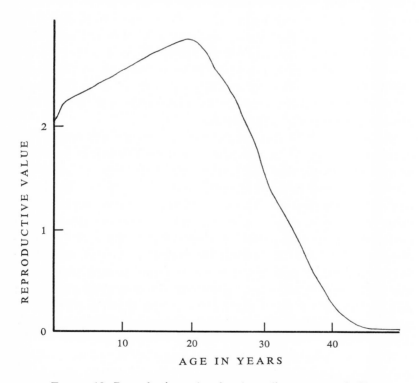

FIGURE 18. Reproductive value for Australian women of different ages. Calculated from the birth and death rates current in Australia about 1911. (From Fisher, 1958)

tion. The early years are below the peak, because besides reproductive potential the chance of survival must be included. However after the peak the decline in the reproductive value and in the contribution to evolutionary change sets in rapidly. In man, for instance, they are both almost negligible after forty years of age. This means that from the point of view of selection, older individuals make no contribution and therefore it is easy to imagine that their elimination would be adaptive. For after all, if they make no contribution to the subsequent generations, their presence is no more than a drag on the food supply of the population.

This statement neglects the fact that the post-reproductive period may be adaptive for other reasons. Older individuals could be repositories of knowledge and experience of great value to the population, especially in relatively social animals. However, as Fisher (1958) points out, these advantages would also have a positive effect on the over-all reproductive value and individual survival of a population, and therefore there would be some selective pressure to retain some older individuals in the population. If the individual is making a contribution to the population either by reproduction or by wisdom, he will have in both cases some degree of adaptiveness.

Medawar (1957) has argued that a possible advantage of a declining old age is in the elimination of deleterious genes. It would seem to me that this is only one of the subordinate benefits which accrue from eliminating individuals of low reproductive value. Because of the shape of the curve of reproductive value, any mechanism, any mutation or modifying effect which moves the deleterious action to a period after the peak in reproductive value, will be favored by selection. In this way there will be a general pushing of harmful genes toward the end of the cycle, making old age a graveyard for unwanted genes. But this can be only a small part of the reason that the life span is adaptive, the principal one being the direct relation to the reproductive value.

A second possible adaptive importance of life spans concerns the degree of recombination and the control of the amount of variation, a subject already discussed in detail. The duration of the life span and generation time are factors which affect the rate of recombination. However, it would be hard to say that a particular length of life span has evolved in order to produce a given degree of variation control, for it may also be true that the rate of recombination adjusts itself by natural selection to a given life span. All we can really say is that the two are correlated, and one will affect the other during evolution. One could say crudely that if generations were not cut down to the minimum duration possible for a given size, the rate of evolution (and recombination) would be unnecessarily delayed. In fact there will be a positive selection pressure to shorten generation time so as to permit a higher degree of recombination. Generation time is directly related to life span, especially if one considers it in terms of reproduction value (Fig. 18).

In discussing the effects of generation time in evolution, Wynne-Edwards (1962) makes the point that while the short cycle increases the frequency of possible gene recombinations, the long cycle makes for stability in numbers. It does this by mixing the breeders of different ages or year classes, by permitting the parents to be associated with the offspring for longer periods and therefore supervise training, and by increasing the opportunities of tradition and convention, which he believes are so vital to number control. He further points out that the most successful groups of animals on the surface of the earth—insects, passerine birds, and rodents—all tend to have relatively short life spans as compared to their close relatives, which suggest that the advantages of the short cycle have contributed to the success of these groups.

A third possible adaptive significance of life spans concerns their relation to size. This was considered by Bidder (1932), who argued that size limits are the prime factor of adaptive significance, and that the duration of the life span and senescence are merely by-products. He pointed out that size limits are advantageous only to certain types of organisms such as terrestrial animals, which are subject to mechanical problems if they become too large. This is not the case for some immobile plants, or for aquatic organisms which are buoyed by the water, and indeed these organisms often do not have size limits but grow indefinitely, and furthermore they fail to exhibit senescence. For example, these conditions were found in giant trees, tissue culture of cells, some fish, and in various invertebrates such as sponges and corals.

There are surely other reasons for the adaptiveness of specific life spans, and it is hardly necessary to point out that these could act together or in different combinations for different organisms. Having considered the reasons why one might have definite life spans, it may be helpful to examine the ways in which these spans are achieved. What are the principal means of eliminating individuals? They may be subdivided into three main categories.

The first may be called ecological death. This is the simplest situation; the rigors of the external world, competition with rivals, weather, disease, and all the other hardships of the environment winnow the population. It need not involve some sudden catastrophe, a hurricane or a flood; it may be simply the normal steady-state condition, fully compensated for by the rate of reproduction. There is considerable evidence, from comparing the life spans of individuals of the same species in nature and in captivity, that this is a strong factor in keeping the life span of many organisms short.

The second method of eliminating individuals is an internal rather than an external one. Some species have a genetically controlled mechanism causing them to die suddenly. This is sometimes called "beneficial death," and the most striking examples come from the insects (see Emerson, 1960). Mayflies, for instance, will undergo a long slow larval period in stream beds, pass through an elaborate pupation, and produce a wholly new and larger imagoes which hatches from the water in great swarms (to the joy of the trout and the trout fisherman). These adults survive only a few hours, just long enough to accomplish fertilization and oviposition, and then they die. In a similar way the annual plant does much the same. A cereal such as wheat (Plate 17) goes through a period of growth, and shortly after fertilization and seed formation are accomplished the whole plant, except for the seed, turns golden and dies. These cases are "beneficial" in the sense that since the reproductive value of the individual has suddenly fallen to zero, it is clearly advantageous that it be eliminated and no longer be able to compete with individuals whose reproductive value is high.

The third and last mechanism of elimination (also internal, like the second) is by senescence, which is a slow and steady decline in the functioning of an organism. We have already examined the mechanisms whereby this is achieved, and it might be added here that this is similar to the sudden, beneficial death described above, except that the process is long drawn out. One can easily see that ecological death can be more effective if it operates on senescent individuals; indeed the two factors must operate together.

The period of size equilibrium

In considering the matter of the adaptive advantages of size equilibrium or dormancy we must first briefly reiterate some points already made. Resistance by dormancy is of obvious advantage for weathering unfriendly environmental circumstances, and this is confirmed by the numerous ways in which the entering of the dormant period is triggered by environmental events (e.g., starvation, day- and night-length recognition, etc.). And the end of dormancy is equally dependent upon stimuli from the outside (e.g., periods of cold followed by warmth).

There is also an interesting correlation between a resistant stage and recombination. Since many of the spores are either zygotes (zygospores) or gametes, the products that follow upon germination will be recombined and therefore correspondingly variable. The advantage of such a system is obvious: there is a high probability that the emerging spore products will face new environments, and the greater the genetic variability, the greater the likelihood of having a genotype well-suited to the new external conditions. A few examples will illustrate the point.

The unicellular alga *Chlamydomonas* is a good case of an organism in which the vegetative stage is haploid (Plate 3). The vegetative individuals may divide into two, four, or eight daughter cells within the parent wall, and depending upon the species they may be liberated immediately or after a period in a non-motile palmella stage. This asexual reproduction in no way involves a resistant stage. At the onset of sexual reproduction the individual cells function as gametes, and the zygotes form a thick wall and pass through a period of rest. Upon germination the zygote undergoes meiosis and produces four or more flagellated daughter cells. The genetic variability resulting from recombination will be present in these individuals that emerge from the resistant zygotes.

Volvox is a more advanced relative of *Chlamydomonas*, and the situation is very much the same except that the vegetative individual consists of many hundreds of cells bound together in a colony (Plate 9). The asexual reproduction is achieved by certain cells in the colony that divide and produce a daughter colony which forms as an inward pocket and escapes upon the disintegration of the mother colony. As pointed out earlier, asexual reproduction continues throughout the summer when conditions are fairly constant and favorable for growth, but as soon as winter begins the colonies produce gametes instead of daughter colonies. The small sperm fertilizes the large egg, and the zygote rapidly develops a hard spiny wall. This resistant cyst will settle in the mud and survive the winter; in the first warmth of spring it will swell, and following meiosis and successive divisions it will produce a new small colony that will subsequently produce progeny by asexual budding. However, it is important to note that all the products of meiosis are kept in the one initial colony (see Papazian, 1954). This means, therefore, that only part of the reshuffling of recombination lies in the resistant spore; two different genomes of two parents are brought together, followed by meiosis, but all the progeny of meiosis are imprisoned in the first colony. This colonial genetic heterogeneity lasts only one generation, since the asexual daughter colonies come from single cells and therefore the subsequent colonies are genetically pure cell clones. The principle that recombination is essentially carried in the resistant stage holds, although the life history of this unusual organism requires an asexual generation before the recombined cell progeny can be completely segregated.

One final example of a haploid organism is the bread mold *Rhizopus*. In heterothallic forms the haploid hyphae of opposite sexes abut against each other at their growing tips. On each side numerous nuclei are cut off by a cross-wall partition, and ultimately these gametangia fuse by dissolving the wall that separates them. Many of the nuclei from both parents now fuse, those that fail to pair degenerate, and at the same time the whole zygospore forms a thick horny coat to form a resistant, resting stage. Meiosis apparently occurs as the zygospore germinates, and generally the first mycelium is very short, producing immediately an asexual sporangium. In this case the added feature is a secondary resistant spore which is found in the sporangium. At first glance it would appear that the rule concerning the association of recombination and spore formation only applied to the zygospore, but the production of asexual spores is part of the scheme as well. As with *Volvox*, immediately after the germination of the zygospore, the genetically different nuclei resulting from meiosis cannot be segregated, since they are held in a common mycelium. Theoretically they could be separated by individual nuclei forming a new hyphal branch and in this way starting a pure nuclear clone. But even more effective is the isolation of nuclei in small groups in the spores (two to ten nuclei per spore in *Rhizopus*) so that the genetic segregation is completed on a large scale in the first sporangium following the sexual events. Thousands of spores will be produced, each with different combinations of the different nuclei, and some of these different nuclear groups have an excellent chance of finding an environment that especially suits their particular genetic constitution.

If one looks elsewhere among the phycomycetes, the water mold *Allomyces* provides an excellent illustration. In the forms which Emerson (1941, 1954) has lumped under the type Euallomyces, there is an alteration of generations in which the sporophyte and gametophyte are both large plants of equal importance (Plate 13). The haploid mycelium produces both male and female gametangia and the uniflagellate gametes fuse to produce a diploid mycelium. At maturity this sporophyte has two kinds of sporangia: a thin-walled type that produces asexual zoospores allowing repeated asexual generations if the conditions are favorable, and a thick-walled resistant sporangium which contains numerous nuclei that divide meiotically at germination before they are cut off into separate uniflagellate swarmers (Wilson, 1952). Each of these haploid swarmers is capable of giving rise to a new gametophyte; therefore there is an immediate segregation of the products of meiosis. But the interesting features are that the sporophyte can produce two different kinds of sporangia and that the sporangia in which recombination occurs are the thick-walled resistant type. Emerson showed that if the latter were germinated precociously, meiosis did not occur and sporophytes were produced instead of gametophytes. There appears to be a close relation between the two types of spores, but in keeping with the rule set forth here, recombination is associated with resistance.

It might be added that these facts apply equally well to the more advanced cystogenes types of *Allomyces*, where the gametophyte has been greatly reduced to small circular protoplasts that produce gametes directly.

The principle may be readily illustrated in the higher fungi also. In Ascomycetes there is recombination and segregation with the formation and dissemination of the resistant ascospores.

The fact that mosses, ferns, and the higher groups of plants also show the correlation is so obvious that the briefest summary is adequate. In the mosses the main body of the plant is haploid, fertilization takes place at the tip, and the sporophyte grows and produces spores at its tip (Plate 15). Since the spores give rise to new gametophytes, sporulation and combination occur together. In ferns the main plant is diploid, the gametophyte being confined to a small delicate prothallus (Plate 16). The prothallus produces gametes which upon fertilization give rise to the fern proper, and all along the edge of the leaves there are spore-bearing sori. The spores, as in the mosses, give rise to the gametophyte, again an association of recombination and sporulation. Beyond the ferns the gametophyte becomes so inconsequential that it eventually is never separated from the great sporophyte, and we have a new innovation which bears on our argument: the substitution of pollen for a motile sperm. Pollen does not possess a very thick wall; nevertheless it is to some extent resistant. Furthermore, it certainly plays an important role in recombination, although a partially different role from that of our previous examples. Pollen carries the products of meiosis of one of the two gametes; therefore it does not directly benefit the whole organism in producing resistant variants to face new environments. Instead the benefits are confined entirely to fertilization; it is an effective method of permitting variable male gametes to travel great distances without destruction by desiccation and other environmental severities.

In metazoa the fertilized egg in some cases is resistant and therefore an exact equivalent to the zygospore. When it is of the resistant type, its relation to recombination is again simple and straightforward. Different animals vary as to how much development has occurred before the egg becomes hardened; in some cases there are simply the male and female pronuclei, while in others an embryo will be partially formed. In this, as well as in the storing of food in the form of yolk along with the embryo, metazoa resemble the higher plants.

A striking illustration of the correlation occurs among certain aphids. In numerous species of these insects the sole method of reproduction during the summer is through the viviparous production of parthenogenetic individuals. By this particular mode of specialized asexual budding, the aphids produce many offspring that may immediately benefit from the plant juices that are abundant in the warm seasons of the year. As the cold weather approaches sexual males and females appear, fertilization ensues and resistant eggs that survive the winter are deposited externally on the surface of the host plant. Therefore the only time that meiosis occurs in the annual cycle of these aphids is also the only time that resistant eggs are produced, and these hatch with the new season the following year.

All these examples show recombination produced by sexual reproduction, but it will be remembered that recombination is also possible in a modest degree in asexual forms, especially heterokaryotic Ascomycetes. And again in this case the new combinations of nuclei are wrapped up in the macroconidia; recombination is, as before, associated with a resistant stage.

Conclusion

Life cycles are made up of steps and interweaving chains of steps, all of which could theoretically be adaptive and many of them are. For instance, if one examines the various parts of the life cycle, it is possible to dissect out some especially striking adaptive advantages of certain stages. This is obvious for the period of size increase, which contains the lion's share of the steps, particularly if one classifies the steps according to their function, their division of labor. In the period of size decrease abrupt changes provide the separation necessary for cycling to occur and also produce the cell units required for recombination by meiosis and fertilization. The adaptive advantage of the period of gradual size decrease, more especially senescence, is intimately connected, at least in some organisms, with the control of the life span, which is also affected by the degree of recombination and the evolutionary stability of an organism. The point of minimum size in the cycle serves both the adaptive function of carrying the organism through adverse periods in unfavorable environments and the function of timing recombination so that it occurs at the most effective moment in the cycle of the organism in relation to the environment.

It is no coincidence that so much of the life history is deeply involved with recombination. This is one of the key factors that govern adaptive change in periods of decrease and equilibrium. There are other factors also, some of which operate at the same time and may be interrelated by working in the same or opposite directions. Furthermore, all these changes helping the organism to take advantage of ecological opportunities are sequences of steps that are ultimately gene controlled. The various chains may be interrelated to one another either at the genetic level or at the level of one or more of the steps. Therefore the life cycle may be thought of as a complex of chains of steps being continually encouraged and discouraged by a complex of environmental opportunities. The result is evolution, but the fact that many of the steps at both ends of the chains are connected makes isolating any one portion a difficult and certainly an artificial task.

THE NUMBER OF STEPS

Size increase in evolution: the facts

Without a moment's reflection any biologist would agree that size changes have occurred during the course of evolution, but if one made the sweeping statement that there has been an overall increase in size during the course of geological history, this would likely be contested. Yet in a very rough sense it can be demonstrated very easily. The point to remember is that the maximum size possible is the thing that has increased since the

origin of life. This does not exclude the possibility (or the un-
doubted probability) that there have been size decreases also,
but the maximum and perhaps the mean size show a fairly steady
increase.

This notion can be tested in a crude fashion by finding the
maximum length of various groups of organisms known from
the paleontological record for different geological periods. If
these are plotted against a time scale, one can see from Figure 19
that there is a sweeping increase since the first blue-green algae.

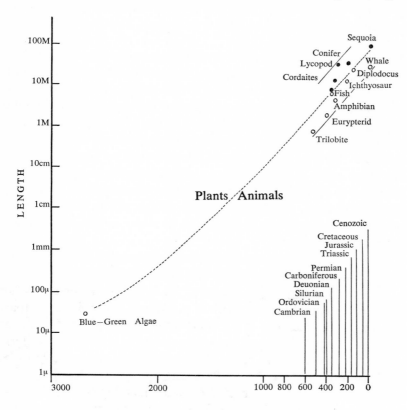

FIGURE 19. Graph showing the maximum sizes (shown as the
logarithm of the length or height) of animals and plants from
different geological eras (plotted linearly). (The author is
grateful to Drs. Baird, Dorf, Fischer, and Jepson of the Geology
Department, Princeton University, for their help in obtaining
the data for this graph.)

Unfortunately there is a large gap between these algae and the cambrian trilobites. But even before the algae we might assume that the points were on the same curve, if the postulate that bacterial cells preceded blue-green algae is correct.

From the cambrian to the present day the points are rather irregular, yet there undoubtedly is an over-all upward trend. It is helpful to distinguish plants from animals, for within each kingdom the upward trend is more obvious. Since this curve is for maximum sizes, it not only conceals the fact that there have been size decreases, but also that certain organisms such as bacteria and blue-green algae have not changed their size for billions of years.

Besides these slow long-term trends upward of maximum size, there are some most interesting, relatively short-term upward trends among numerous animal fossil series. These were first described by Cope (1885, 1896) for vertebrates, and in fact the phenomenon is often referred to as Cope's law. Newell (1949) has made an excellent survey of similar trends among invertebrates. A number of these have been plotted on Figure 20 with the same axes as Figure 19 to show how these examples of Cope's law differ from the increase in the maximum described previously: these trends occur over comparatively short periods of geological time and their rate of increase is far greater than the rate of the maxima.

If we now turn to size decrease, Newell makes the very interesting point that among the fossil series there are none that show a corresponding gradual decrease; that is, the reverse of the many cases of gradual increase is not found. He points out that this might be a defect in the paleontological record, for there are many groups of small organisms such as mites and rotifers which we assume came from larger ancestors and have left no fossil record.

There is one interesting case of extremely short-term size decrease that should be included here. This is the case of diatoms, where the peculiarities of their silicon shell impose progressive size reduction as a result of successive fissions. The valves are like the bottom and the lid of a pill box, and following mitosis the two valves separate, each containing one of the daughter nuclei. They now proceed to synthesize a new valve on the exposed

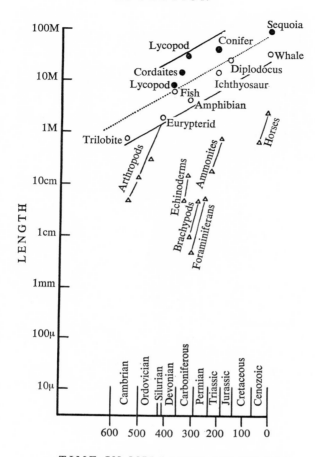

FIGURE 20. Cope's law illustrated by a graph showing the relatively short-term trends of size increase in certain animals. These have been plotted on a portion of the graph shown in Figure 19 so that the two rates of increase may be compared. (Data from Newell, 1949)

half, and this new half always forms inside (that is, a new bottom) of the old valve, regardless of whether the old one was a top or a bottom. The result is that in some species the mean size of the individuals decreases steadily with each division (Fig. 21). It is the structure of the valve that causes this phenomenon, and

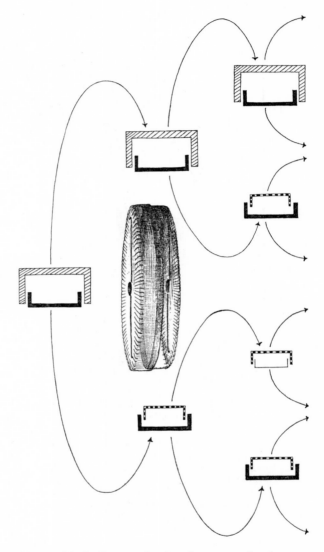

FIGURE 21. A diagram showing the progressive decrease in the size of the valves of a diatom. The central drawing shows how the valves fit together, and since after cell division each valve develops a new "bottom" which fits inside, the mean size of the population decreases. The size diminution in the diagram has been greatly exaggerated to illustrate the point.

obviously some antidote is necessary, for otherwise all members of these particular species of diatoms would soon disappear from sight. The problem is solved by periodic auxospore formation, which may be a result of sexual conjugation or of autogamy; but in either case a totally new large shell is formed. Nipkow (1927) has made a study of the deposits on the bottom of Swiss lakes and has come to the conclusion that the time between auxospore formations, that is the period of gradual cell-size reduction of the clone, must be in the order of two to five years. This is a mere instant in time if it is compared to the examples of Cope's law, and from our point of view it is no more than a curiosity, for clearly this is not an adaptive size change but a mechanical necessity.

It would be possible to go into far greater detail on the evidence for size changes during the course of evolution, but for our purposes this is hardly necessary. All that needs to be demonstrated is that they do occur, and what are the general characteristics. It is then possible to ask why they have occurred, and what is adaptive about size, or the increase (or decrease) in the number of steps.

The adaptiveness of size

Every aspect of a living organism is so intimately connected with so many other aspects that it is often impossibly difficult to isolate one and show that this particular one is adaptively significant. In the first place adaptation is an abstraction that is impossible to test, and in the second place more than one of the interrelated aspects might be selectively advantageous. Or some might be, and others disadvantageous, but all that we can see is the over-all value of the complex.

It is therefore more important to try to find the interrelations between the number of steps and related phenomena, rather than merely make guesses about adaptive significances. If we can see how one change affects others, this surely is a beginning towards understanding how the whole has been successful in evolution.

If we begin from the question of size or the number of steps during the period of size increase, we can see that as size varies, the following are affected: (i) the amount of variation, (ii) the ecological relations of the organism, (iii) the complexity of the organism, and finally, (iv) the stability of the organism. Clearly, each of these will play a significant part in the adaptive success of the organism, which will depend in each case upon the environmental opportunities available at the time and place. These opportunities will change with such factors as time, season, and geographic region, and therefore any simple formula as to what is advantageous and what is not can have little or no meaning. There are many reasons why a change in the number of steps could be good or bad from the point of view of selection, and here we can only expose some of the possibilities or variables in the system.

Size and the amount of variation

This topic has already been discussed in some detail. In the very beginning we showed that generation time varies with size (Fig. 1), and some of the implications concerning variation were discussed in a previous section in this chapter. In particular, we emphasized Wynne-Edwards' (1962) point that the most abundant groups of animals on earth today have comparatively short generation times, which implies advantages in this characteristic.

Generation times have been considered from the point of view, of rates of evolution, especially by Simpson (1953a). His conclusion, based upon paleontological evidence, is unequivocal: there is no correlation between generation time and rate of evolution. He gives many examples to prove the point; for instance, opossums evolved more slowly than elephants; fast breeding cricetines did not change any faster than the slow breeding carnivores and ungulates in the great invasion of South America; Foraminifera have hardly changed at all, yet they are small and have short generations. In fact invertebrates in general are smaller than vertebrates, yet they have evolved more slowly. Stebbins (1950) has shown the same to be true for plants, for some herbaceous plants evolved more rapidly than woody plants, and others more slowly.

The difficulty with this kind of information is that by necessity it consists of isolated examples. More significant would be a plot of evolutionary rates against size for a large number of organisms, to see if there is an over-all correlation or whether the relation is largely random. Also, it must be clear that rates of evolution are an entirely different matter from abundance at any one moment in geological history.

Unfortunately we can conclude at the moment only that the effect of size on the amount of variation (expressed either in terms of abundance at any one epoch or in rates of evolution) is unclear. Rationally, we know that the frequency of shuffling (which is the frequency of reproduction or reciprocal of the generation time, which is correlated with size) must affect the amount of over-all variation and rates of change. But clearly we have insufficient information at the moment to show precisely what the effect is. Furthermore, the effect may be so small that it is negligible and is subordinated by other, more powerful effects.

Size and ecological relations

The fact that the size of an organism has ecological ramifications and implications has been recognized by biologists for many years. This was discussed by Wallace in his essay submitted to the Linnaean Society of London along with Darwin's in 1858, and the main outlines of the argument were set forth with simplicity and clarity by Elton (1927) in his discussion of food chains and the size of food. In general the predator is larger than the prey, and if this is not the case there are usually some compensating mechanisms such as unusually aggressive behavior to make up the deficiency. For any one animal there is a food of optimum size, and since the optimum is smaller than the consumer, the animals in a food chain will be progressively larger as one proceeds from each predator to its predator. To show just how important size can be in feeding, Elton says, "There lives in the forests round Lake Victoria a kind of toad which is able to adjust its size to the needs of the moment. When attacked by a certain snake the toad swells itself out and becomes puffed up to such an extent that the snake is quite unable to cope with it, and the toad thus achieves its object, unlike the frog in Aesop's fable."

The point is even more obvious if one considers radical size differences. For instance, spiders cannot capture cows in their webs, nor can an ant catch a lion. But Hutchinson (1959) makes the interesting point that organisms feed as they develop, and therefore they go through a whole series of size stages during their period of size increase. For instance, a large fish starts as a small fry and the food that it can eat must correspondingly go through a great size range during the course of time. In fact many fish are cannibalistic and eat their own young when they happen to be of the right size. A very interesting situation arises in various ciliate protozoa (e.g. *Oxytricha*, Giese, 1938) where under conditions of mild starvation a clone will become bimodal in size distribution; some of the individuals turn cannibal and become relatively enormous, while the others become prey and remain relatively small. Any one of these individuals, either large or small, if cloned in a nutritious medium, will give rise to a standard size population. These size differences are entirely phenotypic, and they represent a special case of range variation.

Another aspect of size relations related to feeding is the speed of locomotion. The predator must catch the prey and the prey must escape the predator. This matter is one of considerable confusion, for one often hears that size gives advantage in velocity, although those who have discussed the matter in detail do not appear to agree. For instance, A. V. Hill (1950) has argued that in similarly constructed organisms one would expect the speed to be the same, the larger animal having an advantage only in staying power (by being allowed a larger oxygen debt). There are so many factors that can affect speed that, rather than choose two organisms which are closely related and of small size difference, I have considered the greatest possible range of sizes. For these the maximum (reliable) speed was obtained and plotted against the length. This was done for organisms that swim, run, and fly (Table 2; Fig. 22).

There are a number of surprising features to this graph. In the first place the general statement that large organisms can move more rapidly than smaller ones is certainly substantiated. Secondly, if the length of the animal is one millimeter, then running is possibly slightly more effective than swimming, but flying is vastly faster than either running and swimming. On the other hand if the organism is approximately one meter long, then its speed is roughly the same whether it runs, swims, or flies. The

Table 2. The maximum speed of organisms of different size. The table is subdivided into 3 types of locomotion: swimming, running, and flying. This table is the basis for Figure 22.

Species	Length		Speed in cm/sec	Reference
	Swimming			
1. *Bacillus subtilus*	2.5	μ	1.5×10^{-3}	Tabulae Biologicae
2. *Spirillum volutans*	13	μ	1.1×10^{-2}	idem
3. *Euglena* sp.	38	μ	2.3×10^{-2}	idem
4. *Paramecium* sp.	220	μ	1×10^{-1}	idem
5. *Unionicola ypsilophorus* (water mite)	1.3	mm	4×10^{-1}	Welsh (1932, J. Gen. Physiol. *16*: 349)
6. *Pleuronectes platessa* (plaice; larva)	7.6	mm	6.4	Boyar (1961, Trans. Amer. Fish. Soc. *90*:21)
7. idem	9.5	mm	11.5	idem
8. *Carassius auratus* (Goldfish)	7	cm	75	Bainbridge (1961, Sympos. Zool. Soc. London *5*:13)
9. *Leuciscus leuciscus* (dace)	10	cm	130	idem
10. idem	15	cm	175	idem
11. idem	20	cm	220	idem
12. *Pomolobus pseudoharengus* (river herring)	30	cm	440	Dow (1962, J. Conseil Intern. Explor. de la mer *27*:77)
13. *Pygoscelis adeliae* (Adelie penguin)	75	cm	380	Meinertzhagen (1955, Ibis *97*:81)
14. *Thunnus albacares* (yellowfin tuna)	98	cm	2080	Walters and Firestone (1964, Nature *202*: 208)
15. *Acanthocybium solandri* (wahoo)	1.1	M	2150	idem
16. *Delphinus delphis* (dolphin)	2.2	M	1030	Hill (1950, Sci. Progress, *38*:209)
17. *Balaenoptera musculus* (blue whale)	26	M	1030	idem

Species	Length	Running Speed in cm/sec	Reference
1. *Bryobia* sp. (clover mite)	0.8 mm	8.5×10^{-1}	Pillai, Nelson, and Winston (Pers. comm.)
2. Species of Anyestidae (mite)	1.3 mm	4.3	idem
3. *Iridomyrmex humilis* (Argentine ant)	2.4 mm	4.4	Shapley (1920 PNAS, 6:204; 1924, *10*:436)
4. *Liometopum apiculatum* (ant)	4.2 mm	6.5	idem
5. *Peromyscus M. bairdii* (deermouse)	9 cm	250	Layne and Benton, (1954, J. Mammal. *35*:103)
6. *Callisaurus draconoides* (zebra tailed lizard)	15 cm	720	Belkin (1961 Copeia, p. 223)
7. *Tamias striatus lysterii* (chipmunk)	16 cm	480	Layne and Benton (1954, J. Mammal. *35*:103)
8. *Diposaurus dorsalis* (Desert crested lizard)	24 cm	730	Belkin (1961, Copeia, p. 223)
9. *Sciurus carolinensis leucotis* (grey squirrel)	25 cm	760	Layne and Benton (1954, J. Mammal. *35*:103)
10. *Vulpes fulva* (red fox)	60 cm	2000	Hill (1950, Sci. Progress, *38*:209)
11. *Acinonyx jubatus jubatus* (cheetah)	1.2 M	2900	idem
12. *Struthio camelus* (ostrich)	2.1 M	2300	idem

(Note: In this table the length of the quadrupeds is estimated in two different ways. With mammals it is from the base of the tail to the tip of the snout. In lizards one half the tail is also included. In the ostrich it is the erect height.)

Flying

Species	Length	speed in cm/sec	Reference
1. *Drosophila melanogaster* (fruit fly)	2 mm	190	Hocking (1953, Trans. Roy. Ent. Soc. *104*:223)
2. *Tabanus affinis* (horse fly)	1.3 cm	660	idem
3. *Archilochus colubris* (ruby-throated hummingbird)	8.1 cm	1120	Pearson (1961, Condor, *63*:506)
4. *Anax* sp. (dragon fly)	8.5 cm	1000	Wigglesworth (1939, Prin. Insect Physiol.)
5. *Epesticus fuscus* (big brown bat)	11 cm	690	Hazard and Davis (1964, J. Mammal. *45*:236)
6. *Phylloscopus trochilus* (willow warbler)	11 cm	1200	Meinertzhagen (1955, Ibis, *97*:81)
7. *Apus apus* (swift)	17 cm	2550	idem
8. *Cypsilurus cyanopterus* (flying fish)	34 cm	1560	Idem and Schultz and Stern (1948, The Ways of Fishes)
9. *Numenius phaeopus* (whimbrel)	41 cm	2320	Meinertzhagen (1955, Ibis, *97*:81)
10. *Anas acuta* (pintail duck)	56 cm	2280	idem
11. *Cygnus bewicki* (Bewick's swan)	1.2 M	1880	idem
12. *Pelicanus onochrotalus* (white pelican)	1.6 M	2280	idem

third point of interest is that in swimming or walking there is a very crude linear relation between length and speed. While in flying the speed is approximately proportionate to the cube root of the length, a fact which Hocking (1953) had previously noted for insects. There are also some points of minor interest. For instance, beyond a certain size limit there seems to be no further increase in speed, and whales are no faster than fish or dolphins. Since flying fish are only a third of a meter long they

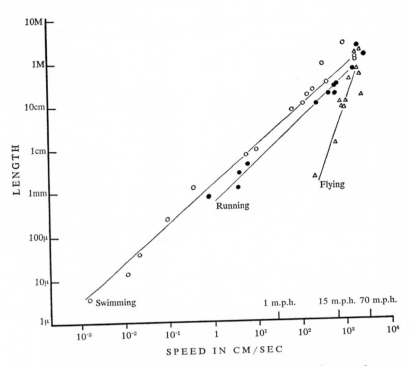

FIGURE 22. The maximum swimming, running, and flying speed of organisms of different size. The length of the animal is plotted against the velocity on a logarithmic scale. The data have been selected for a diversity of types of organisms and for the most rapid examples within each group. It is possible to identify the animals involved by consulting Table 3, which gives the data and the sources for this graph. The Table is grouped according to the type of locomotion and begins with the smallest organism for each category.

clearly derive benefit from speed in escaping their predators, by going into the air. If they were a meter long they would be just as well off if they remained in the sea, which perhaps accounts for the fact that tuna fish do not fly.

The theoretical basis of these relations are difficult matters beyond the scope of the present discussion, although a few brief points should be mentioned. Hill (1950) shows that the rate of contraction of a muscle (or intrinsic speed of contraction) is inversely related to its size, and that therefore the distance gained by size increase, for instance in a step, is lost by the slower rate at which the step is taken. By comparing animals moderately close in size, he shows that their rates of movement are about the same. Here, by taking a much larger size range and not worrying about similarities in construction, we see that all organisms do not move at the same speed, and the size ranges where similar speed is possible are relatively restricted. In discussing the matter with Mr. Arthur M. Young, an aeronautical engineer, he has suggested that an additional factor of even greater importance is power loading. That is, with increased size, the horsepower per unit weight must go up. It is of course well known, as D'Arcy Thompson (1942) and others have pointed out, that the proportion of muscle (and bone) to total weight goes up with increased size because of the principle of similitude, by which weight varies as the linear dimensions cubed and strength as the linear dimensions squared. Therefore there are two factors (at least) to contend with: the efficiency or intrinsic speed of muscles and the power loading. Further, in swimming or flying there is the question of viscous drag of the medium, and this varies greatly with the velocity. In the case of flying, the air drag varies roughly as the square of the velocity, so to overcome it the power required is the product of the drag and the velocity, or

$$\text{Velocity} \propto \left(\frac{\text{power}}{\text{weight}}\right)^{1/3}$$

and since power loading $\left(\frac{\text{power}}{\text{weight}}\right)$ is proportional to size as we have just shown, then

$$\text{Velocity} \propto (\text{Size})^{1/3}$$

While this is approximately the relation shown in Figure 22, it can be only part of the explanation for the slope of the curve, although it is hoped it is the most significant one. It is of interest that Mr. Young has tested his arguments by seeing where aircraft would fall on Figure 22 and finding they fall on an extension of the line of flying animals. (For instance, a Boeing 707 is 145 feet long and does 400 m.p.h. sea level velocity.) In this case the power loading is precisely known, and from it we can conclude that the amount of power per pound in large aircraft is very great, while in a fruit fly, or a swimming *Paramecium* the amount of contractile protein could be very small in proportion to body weight.

Without waiting for a final and thorough analysis of the theoretical reasons for the relationship, we can say without hesitation from the empirical evidence that larger animals move faster. This, therefore is clearly another way in which size can be related to properties of adaptive importance.

Another relation between the size of an organism and its feeding habits has to do with territories. Many animals are territorial and it has been shown that in those forms in which feeding takes place in the territory, the territory is proportioned in accordance with the food needs of the individual and his family. These needs are in turn dependent upon the abundance of food in the whole region and the size of the species of animal, for the larger the organism, the more food is needed. On this point it has been shown by Hutchinson and MacArthur (1959) that the area of the home range varies roughly as the square of the head and body lengths for various mammals.

In a much broader sense the ecological nitches are related to size. It is simply that a large organism is unlikely to compete with a small one for they will live in separate size worlds or nitches. Bacteria and any large mammal (or small one, for that matter) do not eat the same thing or in any way conflict in their respective existence, except if one happens to be a parasite of the other. This, incidentally, may well be a significant reason why there has been a trend during the course of evolution to increase the size of organisms. Since all the smaller nitches will be occupied, the only way to conquer new worlds is to make larger nitches. It is only through catastrophe or some peculiar change of conditions that the smaller nitches will be vacated, and there might be a selection pressure for a reversion to smaller size.

The last ecological consideration of individual size to be mentioned here is again one familiar to ecologists. As the size of the individual increases, on the average the number of such individuals (that is their abundance in nature) is bound to decrease. Elton (1927) has shown that this is connected with the question of the food chain, for on the basis of energy requirements, there must be more protoplasmic bulk at the lower level of small organisms than at the higher level of large organisms in the chain, and the result is what Elton has called the pyramid of numbers.

In a recent study Slobodkin (1964) describes some pertinent experiments on hydra. He has used two species, a green hydra (*Chlorohydra viridissima*) and a brown hydra (*Hydra littoralis*) which differ in body size, the brown hydra being larger. It is presumed that the larger size has the advantage of permitting the capture of larger food and a greater capacity to withstand periods of starvation, while the smaller species can take in more small food more rapidly, and of course its generation time is shorter; therefore the smaller species is admirably equipped for fast growth in periods of abundant small food. In a series of experiments in which the amount and size of the food was varied, Slobodkin was able to show that under some conditions the large and slow growing species thrived, while in others it was advantageous to be small and fast growing.

These results have been effectively interpreted by Slobodkin in terms of strategy, for again one sees that there are multiple components both within the organism and in the external environment. Furthermore these components affect one another in both simple and compound ways. The result, as he emphasizes, is like a game in which the players can never withdraw and stop playing; the whole world is sealed into one great gambling house.

The pertinence of this view can be seen again when another fact is pointed out. If the pyramid of numbers exists, then large animals will tend to exist in smaller populations. Population geneticists, Sewall Wright in particular, have stressed that the amount of variation and the rate of evolution can be affected by population size (review: Dobzhansky, 1941). Of course there are other factors besides the size of the organism that will affect the population size, but this is one of them, and it provides just one more link in the ramifications of the strategy of the game, and one more evidence of the far-reaching importance of size in understanding biological problems.

Size and complexity

We have already discussed the fact that with increased size there is an increase in complexity in the sense of differentiation or division of labor. This was shown by looking at maximum size for a given number of cell-types, and it was obvious that with increased size the minimum number of cell-types increases greatly (Fig. 2). Also, we saw that this was to be expected on the basis of surface-volume considerations; the relationship has been termed the principle of magnitude and division of labor. These are the basic ideas, and no new ones will be added here. However, they can be amplified to some extent in the context of this chapter: in what ways can complexity (which may be related to size) be adaptively significant?

An apparently simple example comes from the study of allometry, which was developed by Huxley (1932). It is, as is well known, a method of comparing the growth rates of two parts of an organism (X and Y) and can be expressed in the simple relation

$$X = bY^k$$

where b and k are constants, k being the ratio of the growth rates. If this is expressed logarithmically, then k is the slope of the straight line on the graph, which makes the method extremely easy and useful for experimental analysis.

This method has been applied by a number of workers (review: Rensch, 1960) to the series of animals which show a rapid progressive size increase over restricted periods of geological time, that is, which follow Cope's rule (Fig. 20). Thus it can be shown that the alterations of certain features, such as the horns, are not caused by any special new mutations, but rather there has been a simple selection for increased size. The growth rates of the parts have not changed, but if the over-all growth of the animal increases, the result will be a considerable alteration of the over-all proportions of the animal. To give a familiar example, antler size bears an allometric relation to body size in deer, as Huxley (1932) showed in his original monograph. The antlers are so very large in the Irish elk because there has been an increase in body size, and the antlers, in order to keep their positive allometric ratio to body size, have become disproportionally large. The same argument applies to the progressive increase in the horns of *Titanotheres* with increased body size (Fig. 23), and there are a number of other examples (review: Simpson, 1953a).

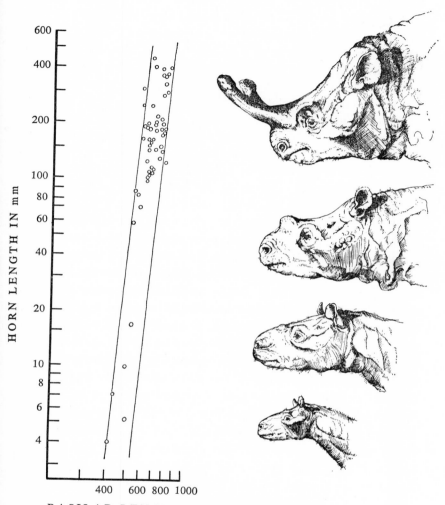

FIGURE 23. Allometry in titanotheres. The horn length is
plotted against the basilar length of a skull on a logarithmic
grid (from Hersch). This relation can also be seen on the ac-
companying drawings of different species of eocene titanotheres
(after Osborn).

The whole matter has been admirably analyzed in great detail by Rensch (1960). He and his co-workers have studied the organs of many species of vertebrates, and his conclusions may best be summarized by his example of "a hypothetical 'rat' of excessive body size, about as large as a beaver," which "would differ from its smaller relatives in the following characters. This animal compared with related smaller species would have a relatively smaller head, brain case (in relation to the facial bones), brain stem (in comparison to the brain as a whole), ears, and feet; a relatively shorter tail and shorter hairs; and a relatively smaller heart, liver, kidneys, pancreas, thyroid, and pituitary and adrenal glands. The weight of the bones would be relatively heavier, the facial bones relatively longer (in relation to the brain case), and the forebrain relatively larger (in relation to the brain as a whole). The retina of this giant rat would be relatively (and probably absolutely) thinner; the layer of ganglion cells and both granular layers of the eye would be less dense; the number of rods and cones would be relatively smaller. In the forebrain the cortex-7-stratificatus would be relatively larger, and the semicortex relatively smaller. The absolutely larger neurons of the brain would be less dense but would have many more dendritic ramifications. There would be equally large but definitely more numerous blood corpuscles and bone and connective tissue cells, and relatively smaller insulin-producing tissue of the pancreas. Finally, the general metabolism (especially the rate of oxygen consumption, breathing, pulse, and blood circulation) would be decreased; the amount of blood sugar would be less, and . . . the onset of maturity would be postponed, the gestation period and average length of individual age would be prolonged, and the animal would be superior in learning ability and in memory."

This shows, in the most striking way, the kinds of problems that arise as a result of size increase with fixed allometric relations. Furthermore it serves to illustrate a point made previously: it is virtually impossible to isolate any one feature of this vast complex and ascribe adaptive significance to it. The only thing we know is that the sum total of the good and the bad for a particular environment in a particular era must be favorable rather than unfavorable if the organism is to exist at all.

Size and stability

Another of the possible adaptive advantages that accompany increased size and increased complexity is stability. This is manifest in a number of different ways which will be briefly described.

In the first place there is a greater stability, in larger individuals, of the internal homeostasis; homeothermy, uterine development, and all the great myriad of physiological mechanisms that so impressed Claude Bernard in his discussions of the constancy of the *milieu interieur*.

A striking way to look at this increase in stability with size is seen in a passage of Simpson (1953a). He points out that organisms that have annual life cycles in a temperate zone (where there is a harsh winter and a warm summer) do not need necessarily to withstand the severe season but can go into diapause or some other protected period of equilibrium. However, if the animal lives more than one season, then it must develop complex physiological mechanisms to tide the organism over the hard periods. Besides insulation and homeothermy, there are other mechanisms to solve the problem of physiological tolerance, such as hibernation and migration. We can add to this the fact that if an organism has a generation time that spans more than the seasons of one year, then it must be larger, for there is a direct correlation between generation time and size. In this case we can state without misgivings that the stability mechanisms which are adaptive and essential for the existence of an organism that spans more than one annual season, are bound to be correlated with the relatively larger size of the organism.

Stability is related to the generation time in another way which is of special interest. Smith (1954, which is based on a concept of Birch, and discussed in detail by Slobodkin, 1961) relates reproduction rates to the harshness of the environment and shows that the ability to produce offspring is inversely related to the ability to withstand the rigors of the outside world.

The core of the idea centers around the relation

$$R_o = e^{rT}$$

Where R_o is the reproductive rate, that is, the sum of the offspring of each female for a given time interval. It is essentially the replacement rate of the population for if $R_o = 1$ then the population precisely replaces itself each generation, while $R_o = 2$ would be the case of binary fission.

r is the rate of increase or rate of compound interest. It is the potential of the organism under optimal conditions for increase.

T is the generation time.

Smith has plotted this relation (Fig. 24) and from the graph

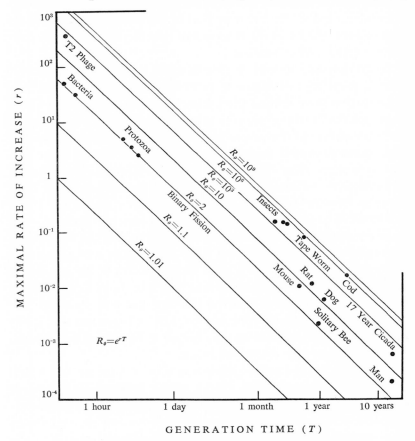

FIGURE 24. The relation between the maximal rate of increase (r), the generation time (T), and the increase per generation or reproductive rate (R_o) for a variety of organisms. The graph is on a logarithmic scale. (After Smith, 1954)

a number of different implications are evident. First of all, it is clear that the generation time is more important than the number of offspring for any one female in permitting a rapid increase in the population. Bacteria with binary fission ($R_o = 2$) are going to increase at a far greater rate than cod which produce millions of eggs ($R_o = 1 \times 10^6$), simply because of their short generation time. We can add to this that since size and generation time are directly related, small organisms can increase at a much greater rate than large ones, even if large ones seem to compensate by having many offspring.

But by far the more important conclusion here is Smith's interpretation of this graph in terms of harshness of environment. He says that "the natural world must become more lenient as a result of a longer period of development . . . man . . . lives in a world 100,000 times more kind than that of his associate *Escherichia coli*. Most of this difference is due to the ability of man, through mere size, through greater complexity, and also through some adaptability, to escape whole classes of environmental hazards that plague the bacteria." Large animals therefore live in a comparatively benign world, benign because they have acquired built-in stability mechanism (through their size-complexity) that can now temper and buffer the vagaries of the environment.

Another related aspect of the stability of the individual through increased size is the effect on the stability of population size. If the environment is variable and the individual small, then the numbers of that individual will climb or crash with the slightest perturbations of external change. However, if the individuals in a population are large, they cease to fluctuate not only because their internal physiological stability helps them to ride with the storms, but also because they could develop more advanced mechanisms of population homeostasis. Therefore the possibilities of adaptive stability exist both at the population as well as the individual level, and in both cases they are related to size— the size of the individuals in the population, and the size (or number of individuals) within the population.

Conclusion

There can be only one conclusion from this chapter, and that conclusion has already been given more than once. Every aspect of an organism, in any part of its life cycle, can be positive, negative, or neutral in its adaptive value; and furthermore this can change with environmental changes (e.g. geographic shifts, climatic changes, changes in the existing fauna and flora). Moreover, all these qualities that are subject to selection can be correlated one with another. For instance, there are correlations between generation time, size, complexity, stability, harshness of the environment, reproductive rates, speed of locomotory animals, prey-predator relations, and undoubtedly many others. Or consider another kind of relation where the aspects or features subject to selection are related to one another by being in a sequence of steps within the life cycle. The interconnections are so prolific that it is difficult if not impossible to isolate any one adaptive feature and explain why it is adaptive. It is too much part of a complex for any abstract dissection. Rather we must be content to see the correlations and know that the sum total must be adaptive.

6. Conclusion

IN the beginning we said that biology was in need of connecting laws that would bring together molecular events, developmental (or life history) events, and evolutionary events. It was pointed out that what would be provided here could hardly be called a law, for in no way can it be formulated with the mathematical precision of physical laws. However, a method has been devised to show how the three kinds of events are connected, how they relate to one another, how they hook together.

To recapitulate briefly, the method involves considering organisms as life cycles, and each life cycle is made up of a series of chemical reactions that have been termed steps. The steps occur in an organized sequence, and they are often grouped together in chains which are to varying degrees dissociable from one another.

The life cycle has a point of minimum size and a point of maximum size. The point of minimum size is the cell (either a zygote or a spore), and this furthermore is the minimum connection, the minimum unit of inheritance from one life cycle to the next. The process of reaching the point of maximum size is usually referred to as development, and it has the largest multitude of steps and chains of steps in the whole cycle.

The point of minimum size is also the point where innovations or variations are introduced, while the point of maximum size is the point where the organism becomes capable of reproduction, and if this is prevented the elimination of variants is achieved. Since natural selection involves both the introduction of new variations and the constant culling of different variants, the life cycle is in this sense the unit of evolution.

Because the life cycle is one complete set of steps (before the repetition occurs in the next cycle) and because it is the unit which permits natural selection to occur, it is the connecting device, the hook, that joins molecular and evolutionary events.

The steps and the chains of steps form complex interweaving sequences which are a meshwork of control systems. The elucidation of these sequences is the urgent task of the biochemist, the biochemical geneticist, and the developmental biologist. They ask how the steps follow one another and how they control one another. It is too much to expect that all the steps of any one life cycle will be known, but it is important to know the general categories of processes and the general systems of feedback and control.

All the steps are capable of being adaptive; they are in fact positively adaptive, negative, or neutral. The adaptive value of all the steps of the life cycle of a particular organism are summed, and if the end result is positive the organism is sustained; if it is negative the organism is eliminated by natural selection. Therefore the adaptive value of the organism depends upon the particular complex of steps within it. Furthermore it depends upon the environmental conditions, which determine essentially whether it is to be preserved or eliminated; the environment provides the value judgments. There are numerous elements in the environment (climate, other organisms, locations, etc.), and these can affect different steps in different parts of the life cycle in different ways. The elucidation of these interrelations is the urgent task of the ecologist and the evolutionary biologist. They are asking why certain steps are favored and others not; what are the conditions that lead to ecological and evolutionary success or failure?

These two great problems, those of the evolution and the composition of life cycles, are similar in that both are centered around the complex interweaving meshwork of steps; they are different in that in one case one wonders how they mechanically operate, and in the other one wonders why they exist, why they are adaptive. Therefore through the study of the life cycle we can narrow the problems of the evolutionary and the molecular biologist down to the steps. Moreover, the factors that affect the steps are of specific concern to all kinds of biologists.

These factors have been in large measure the subject of this book. Their prime characteristic is not only that each one affects the steps but that they affect each other; they are interrelated and part of one fabric. They have been examined systematically in previous chapters, and here we can simply lump the principal factors in a brief list:

size
generation time

complexity (division of labor)
stability
control of amount of variation

Any rigid classification of these factors is somewhat difficult and certainly arbitrary, but there are many minor factors as well which affect the steps of some organisms and are also related to the list above. These are, among a number of others, such factors as:

locomotion (absence or presence; and if present, the speed)
type of cell construction (presence or absence of hard cell wall)
method of responding to environmental cues
reproductive rates
senescence

All these properties might be termed colligative, to borrow a term from the physical chemist. He uses the word to identify the precise relation between vapor pressure, boiling point, freezing point, and the concentration of a solute in a solution; for one cannot vary any one of these factors without altering the others. The only difference between this chemical analogy and the biological colligative properties that have been listed here is that in the former the relation can be expressed mathematically in a simple equation.

If this could also be done for the biological colligative properties, we might really be closer to the kind of connecting laws that we yearn for, and that the physicist is so fortunate in possessing. Since there are so many factors, and the possibilities of interconnections so great, and the number of steps which form their basis so endless, the task seems Herculean. But when we hear of approaches to the problems which utilize concepts of strategy and game theory, as have been suggested for instance by Waddington (1957), Lewontin (1961), and Slobodkin (1964), one sees a glimmer of hope. The population geneticists, especially Fisher, Wright, and Haldane, began the process some years ago, but now it is time to look beyond their restricted task and try to pull all of biology together into one system.

$\mathcal{B}$*ibliography*

Ackert, J. E. 1916. On the effect of selection in Paramecium. *Genetics, 1*:387-405.

Adolph, E. F. 1931. *The Regulation of Size as Illustrated in Unicellular Organisms.* C. C. Thomas, Baltimore.

Alexopoulos, C. J. 1963. The Myxomycetes II. *Bot. Rev., 29*:1-78.

Barth, L. G. 1940. The process of regeneration in hydroids. *Biol. Rev., 15*:405-420.

Beale, G. 1954. *The Genetics of Paramecium aurelia.* Cambridge University Press.

Beermann, W. 1964. Control of differentiation at the chromosomal level. In *Differentiation and Development.* A symposium of the N.Y. Heart Assoc. pp. 49-61. Little, Brown and Co., Boston.

Berrill, N. J. 1961. *Growth, Development and Pattern.* W. H. Freeman, San Francisco.

Bidder, G. P. 1932. Senescence. *Brit. Med. J., 2*:5831.

Blum, H. F. 1962. *Time's Arrow and Evolution.* 2nd ed. Princeton University Press (Harper).

Bonner, James. 1949. Chemical warfare among the plants. *Sci. Amer., 180*:(3)48-51.

Bonner, J. T. 1952. *Morphogenesis.* Princeton University Press (Atheneum).

Bonner, J. T. 1957. A theory of the control of differentiation in the cellular slime molds. *Quart. Rev. Biol., 32*: 232-246.

Bonner, J. T. 1958. *The Evolution of Development.* Cambridge University Press.

Bonner, J. T. 1959a. *The Cellular Slime Molds.* Princeton University Press.

Bonner, J. T. 1959b. Evidence for the sorting out of cells in the development of the cellular slime molds. *Proc. Nat. Acad. Sci., 45*: 379-384.

Bonner, J. T. 1963. Epigenetic development in the cellular slime molds. In *Cell Differentiation.* Ed. by G. E. Fogg (17th Soc. Exp. Biol. Symposium). Cambridge University Press.

Bonner, J. T., A. D. Chiquoine, and M. Q. Kolderie. 1955. A histochemical study of differentiation in the cellular slime molds. *J. Exp. Zool., 130*:133-158.

Bonner, J. T., and M. E. Hoffman. 1963. Evidence for a substance responsible for the spacing pattern of aggregation and fruiting in the cellular slime molds. *J. Embryol. Exp. Morph., 11*: 571-589.

Bonner, J. T., K. K. Kane, and R. H. Levey. 1956. Studies on the mechanism of growth in the common mushroom, *Agaricus campestris. Mycologia, 48*:13-19.

Bonner, J. T., and M. K. Slifkin. 1949. A study of the control of differentiation: the proportions of stalk and spore cells in the slime mold *Dictyostelium discoideum. Amer. J. Bot., 36*: 727-734.

Bower, F. O. 1930. *Size and Form in Plants*. Macmillan, London.

Boyden, A. 1954. The significance of asexual reproduction. *System. Zool., 3*: 26-38.

Braun, W. 1953. *Bacterial Genetics*. Saunders, Phila.

Brien, P., and M. Reniers-Decoen. 1949. La croissance, la blastogénèse, l'ovogénèse chez *Hydra fusca*. (Pallas.) *Bull. Biol. France Belg., 83*:293-386.

Brien, P., and M. Reniers-Decoen. 1952. Apparition de la stolonization chez l'hydre verte, et sa transmissibility. *Biol. Bull. France Belg., 86*:350-380.

Briggs, R., and T. J. King. 1955. Specificity of nuclear function in embryonic development. In *Biological Specificity and Growth*. 12th Growth Symposium, Princeton University Press.

Browning, T. O. 1963. *Animal Populations*. Hutchinson, London.

Bruce, V. G., and C. S. Pittendrigh. 1956. Temperature independence in a unicellular "clock." *Proc. Nat. Acad. Sci., 42*:676-682.

Bünning, E. 1964. *The Physiological Clock*. Academic Press, N.Y.

Buller, A. H. R. 1933. *Researches on Fungi*. Vol. V. Longmans Green, London.

Carpenter, C. R. 1934. A field study of the behavior and social relations of howling monkey (*Aloatta palliata*). *Comp. Psych. Monogr., 10*:1-168.

Castle, W. E. 1929. The embryological basis of size inheritance in the rabbit. *J. Morph. and Physiol., 48*:81-93.

Child, C. M. 1941. *Patterns and Problems of Development*. Chicago University Press.

Cleveland, L. R. 1956. Brief accounts of the sexual cycles of the flagellates of *Cryptocercus. J. Protozool., 3*:161-180.

Comfort, A. 1956. *The Biology of Senescence*. Routledge and Kegan Paul, London.

Commoner, B. 1964. DNA and the chemistry of inheritance. *Amer. Sci., 52*:365-388.

Conklin, E. G. 1905. The organization and cell lineage of the ascidian egg. *J. Acad. Nat. Sci. Phila., 2nd Series, 8*:1-119.

Conklin, E. G. 1931. The development of centrifuged eggs of ascidians. *J. Exp. Zool., 60*:1-119.

Cope, E. D. 1885. On the evolution of the vertebrata. *Amer. Nat., 19*:140-148, 234-247, 341-353.

Cope, E. D. 1896. *The Primary Factors of Organic Evolution.* Open Court Publ., Chicago.

Costello, D. P. 1948. Oöplasmic segregation in relation to differentiation. *Ann. N. Y. Acad. Sci., 49*:663-683.

Costello, D. P. 1961. Larva, Invertebrate. In *The Encyclopedia of the Biological Sciences,* ed. by Peter Gray, pp. 544-549. Reinhold, N.Y.

Crowell, S. 1953. The regression-replacement cycle of hydranths of *Obelia* and *Campanularia. Physiol. Zool., 26*:319-327.

Dalcq, A. 1941. *L'Oeuf et son Dynamisme Organisateur.* Albin Michel, Paris.

Darlington, C. D. 1958. *The Evolution of Genetic Systems.* 2nd ed. Oliver and Boyd, Edinburgh.

Darlington, C. D., and K. Mather. 1950. *Genes, Plants and People.* George Allen and Unwin, London.

de Beer, G. R. 1958. *Embryos and Ancestors.* 3rd ed. Oxford University Press.

Dobzhansky, Th. 1941. *Genetics and the Origin of Species.* Rev. ed. Columbia University Press.

Elton, C. S. 1927. *Animal Ecology.* Macmillan, N.Y.

Emerson, A. F. 1960. The evolution of adaptation in population systems. In *The Evolution of Life,* Vol. 1, ed. by S. Tax. University of Chicago Press.

Emerson, R. E. 1941. An experimental study of the life cycles and taxonomy of *Allomyces. Lloydia, 4*:77-144.

Emerson, R. E. 1954. The biology of water molds. In *Aspects of Synthesis and Order in Growth,* ed. by D. Rudnick, pp. 171-208. Princeton University Press.

Ephrussi, B. 1953. *Nucleo-cytoplasmic Relations in Micro-organisms.* Oxford University Press.

Fauré-Fremiet, E. 1925. *La Cinétique du Développement.* Presses Universitaires de France.

Fauré-Fremiet, E. 1948. Les méchanismes de la morphogenèse chez les ciliés. *Folia Biotheor., Ser. B., 3*:25-58.

Filosa, M. F. 1962. Heterocytosis in cellular slime molds. *Amer. Nat., 96*:79-92.

Fisher, R. A. 1958. *The Genetical Theory of Natural Selection,* 2nd ed. Dover, N.Y.

Garstang, W. 1928. The origin and evolution of larval forms. *Report Brit. Assoc. Adv. Sci. Section D.*, pp. 77-98.

Giese, A. C. 1938. Cannibalism and gigantism in *Blepharisma*. *Trans. Amer. Micr. Soc.*, *57*:245-255.

Goldschmidt, R. 1938. *Physiological Genetics*. McGraw-Hill, N.Y.

Guilcher, Y. 1951. Contribution a l'étude des ciliés gemmipares, Chonotriches et Tentaculifères. *Ann. Sci. Nat. Zool.* (11ᵉ Serie), *13*:33-132.

Gurdon, J. B. 1963. Nuclear transplantation in amphibia and the importance of stable nuclear changes in promoting cellular differentiation. *Quart. Rev. Biol.*, *38*:54-78.

Haldane, J. B. S. 1954. La signalisation animale. *Anné Biol.*, *30*: 89-98.

Haldane, J. B. S. 1956. Time in biology. *Science Progress, 175*: 385-402.

Hansen, H. N. 1938. The dual phenomenon in imperfect fungi. *Mycologia, 30*:442-455.

Hansen, H. N., and R. E. Smith. 1932. The mechanism of variation in imperfect fungi: *Botrytis cinerea. Phytopath, 22*:953-964.

Harper, R. A. 1926. Morphogenesis in *Dictyostelium. Bull. Torrey Bot. Club, 53*:229-268.

Hill, A. V. 1950. The dimensions of animals and their muscular dynamics. *Science Progress, 38*:209-230.

Hocking, B. 1953. The intrinsic range and speed of flight of insects. *Trans. Roy. Ent. Soc. London, 104*:223-345.

Hutchinson, G. E. 1959. Homage to Santa Rosalia or why are there so many kinds of animals? *Amer. Nat., 93*:145-159.

Hutchinson, G. E., and R. H. MacArthur. 1959. A theoretical ecological model of size distributions among species of animals. *Amer. Nat., 93*:117-125.

Huxley, J. S. 1932. *Problems in Relative Growth*. Methuen, London.

Jennings, H. S. 1920. *Life and Death, Heredity and Evolution of the Simplest Organisms*. Gorham Press, Boston.

Kafatos, F. C. 1965. Ph.D. Thesis, Harvard University.

Konigsberg, I. R. 1961. Cellular differentiation in colonies derived from single cell platings of freshly isolated chick embryo muscle cells. *Proc. Nat. Acad. Sci., 47*:1868-1872.

Lack, D. 1954. *The Regulation of Animal Numbers*. Oxford University Press.

Lehrman, D. S. 1959. On the origin of the reproductive behavior cycles in doves. *Trans. N. Y. Acad. Sci., 21*:682-688.

Lehrman, D. S. 1961. Hormonal regulation of parental behavior in birds and infrahuman mammals. Chap. 21 from *Sex and Internal Secretions*, ed. by W. C. Young, pp. 1268-1382. Williams and Wilkins Co., Baltimore.

Lewontin, R. C. 1961. Evolution and the theory of games. *J. Theoret. Biol., 1*:382-403.

Lindaur, M. 1961. *Communication among Social Bees.* Harvard University Press.

Lüscher, M. 1961a. Social control of polymorphism in termites. In *Insect Polymorphism.* Royal Entomological Soc. London.

Lüscher, M. 1961b. Air conditioned termite nests. *Sci. Amer., 205:* (7) 138-145.

Marler, P. 1959. Developments in the studies of animal communication. In *Darwin's Biological Work*, ed. by P. R. Bell, pp. 150-206. Cambridge University Press.

Mayr, E. 1963. *Animal Species and Evolution.* Harvard University Press.

Mazia, D. 1961. Mitosis and the physiology of cell division. In *The Cell*, ed. by J. Brachet and A. E. Mirsky, Vol. 3, pp. 77-412. Academic Press, N.Y.

Medawar, P. 1957. *The Uniqueness of the Individual.* Methuen, London.

Mitchison, J. M. 1957. The growth of single cells. I. *Schizosaccharomyces pombe Exper. Cell Res., 13*:244-262.

Mitchison, J. M. 1958. The growth of single cells. II. *Saccharomyces cerevisieae. Exper. Cell Res., 15*:214-221.

Mitchison, J. M. 1961. The growth of single cells. III. *Streptococcus faecalis. Exper. Cell Res., 22*:208-225.

Morgan, T. H. 1938. The genetic and physiological problems of self-sterility in *Ciona.* I. *J. Exper. Zool., 78*:271-318 (and subsequent papers in the same journal).

Morton, A. G. 1961. The induction of sporulation in mould fungi. *Proc. Roy. Soc. London, Ser. B, 153*:548-573.

Newell, N. D. 1949. Phyletic size increase, an important trend illustrated by fossil invertebrates. *Evolution, 3*:103-124.

Nickerson, W. J., and S. Bartnicki-Garcia. 1964. Biochemical aspects of morphogenesis in algae and fungi. *Ann. Rev. Plant Physiol., 15*:327-344.

Nipkow, H. F. 1927. Über das Verhalten der Skelette planktischer keiselalgen im geschichfeten Tiefenschlamn des Zürich-und Baldegersees. *Rev. Hydrol., 4*:71-120.

Papazian, H. 1954. A theoretical aspect of the genetics of volvox. *Amer. Nat. 88*:172-174.

Pardee, A. B. 1961. Response of enzyme synthesis and activity to environment. *Symposia Soc. Gen. Microbiol., 11*:19-40.

Picken, L. E. R. 1960. *The Organization of Cells and Other Organisms.* Oxford University Press.

Pontecorvo, G. 1958. *Trends in Genetic Analysis.* Columbia University Press.

Prescott, D. M. 1955. Relation between cell growth and cell division I. *Exper. Cell Res., 9*:328-337.

Prescott, D. M. 1956. Relation between cell growth and cell division II. *Exper. Cell Res., 11*:86-98.

Quinlan, M. S., and K. B. Raper. 1964. Development of the Myxobacteria. In *Encyclopedia of plant physiology,* ed. by W. Ruhland, Vol. XV/1. Springer-Verlag.

Raper, J. R. 1954. Life cycles, sexuality, and sexual mechanisms in the fungi. In *Sex in Microorganisms,* ed. by D. H. Wenrich, AAAS (symposium). Washington.

Raper, K. B. 1940. The communal nature of the fruiting process in the Acrasieae. *Amer. J. Bot., 27*:436-448.

Raper, K. B. 1941. Developmental patterns in simple slime molds. *Growth, Symposium. 5*:41-76.

Raper, K. B., and M. S. Quinlan. 1958. *Acytostelium leptosomum*: A unique cellular slime mold with an acellular stalk. *J. Gen. Microbiol., 18*:16-32.

Rasmont, R. 1962. The physiology of gemmulation in fresh-water sponges. In *Regeneration,* ed. by D. Rudnick. 20th Growth Symposium. Ronald Press, N.Y.

Rensch, B. 1960. *Evolution above the Species Level.* Columbia University Press.

Reverberi, G. 1961. The Embryology of Ascidians. *Adv. in Morphogenesis, 1*:55-102.

Rose, S. M. 1939. Embryonic inductions in ascidia. *Biol. Bull., 77*: 216-232.

Ross, I. K. 1960. Studies on diploid strains of *Dictyostelium discoideum. Amer. J. Bot., 47*:54-59.

Rusch, H. P. 1959. In *Biological Organisation: Cellular and Subcellular,* ed. by C. H. Waddington, pp. 263-271. Pergamon Press, London.

Russell, G. K., and J. T. Bonner. 1960. A note on spore germination in the cellular slime mold, *Dictyostelium mucoroides. Bull. Torrey Bot. Club, 87*:187-191.

Samuel, E. W. 1961. Orientation and rate of locomotion of individual amoebae in the life cycle of the cellular slime mold, *Dictyostelium mucoroides. Develop. Biol., 3*:317-335.

Scherbaum, O., and E. Zeuthen. 1954. Induction of synchronous cell division in mass cultures of *Tetrahymena piriformis. Exper. Cell Res., 6*:221-227.

Schmalhausen, I. I. 1949. *Factors of Evolution*. Blakiston, Phila.

Schneirla, T. C., and R. Z. Brown. 1950. Army-ant life and behavior under dry season conditions. (Part 4) *Bull. Amer. Mus. Nat. Hist. 95*:267-353.

Shaffer, B. M. 1961. The cell founding aggregation centers in the slime mold *Polysphondylium violaceum. J. Exp. Biol., 38*:833-849.

Shaffer, B. M. 1962. The Acrasina. *Adv. in Morphogenesis, 2*:109-182; *3*:109-182.

Shaffer, B. M. 1963. Inhibition by existing aggregations of founder differentiation in the cellular slime mold *Polysphondylium violaceum. Exp. Cell. Res., 31*:432-435.

Simpson, G. G. 1953a. *The Major Features of Evolution*. Columbia University Press.

Simpson, G. G. 1953b. The "Baldwin effect." *Evolution, 7*:110-117.

Sinnott, E. W. 1960. *Plant Morphogenesis*. McGraw-Hill, N.Y.

Slobodkin, L. B. 1961. *Growth and Regulation of Animal Populations*. Holt, Rinehart, and Winston, N.Y.

Slobodkin, L. B. 1964. The strategy of evolution. *Amer. Sci., 52*: 342-357.

Smart, R. F. 1937. Influence of certain external factors on spore germination in the Myxomycetes. *Amer. J. Bot., 24*:145-159.

Smith, F. E. 1954. Quantitative aspects of population growth. In *Dynamics of Growth Processes*, ed. by E. J. Boell. 11th Growth Symposium. Princeton University Press.

Sonneborn, T. M. 1957. Breeding systems, reproduction methods and species problems in Protozoa. In *The Species Problem*, ed. by E. Mayr, pp. 155-324. AAAS (symposium) Washington, D.C.

Sonneborn, T. M. 1963. Does preformed cell structure play an essential role in cell heredity? In *The Nature of Biological Diversity*, ed. by J. M. Allen. McGraw-Hill, N.Y.

Sonneborn, T. M., and R. V. Dippell. 1960. Cellular changes with age in Paramecium. In *The Biology of Aging*, ed. by B. I. Strehler *et al.* AIBS, Washington, D.C.

Stebbins, G. L., Jr. 1950. *Variation and Evolution in Plants*. Columbia University Press.

Steward, F. C., E. M. Shantz, J. K. Pollard, M. O. Mapes, and J.

Mitra. 1961. Growth induction in explanted cells and tissues: metabolic and morphogenetic manifestations. In *Synthesis of Molecular and Cellular Structure*, ed. by D. Rudnick. Ronald, N.Y.

Strehler, B. L. 1962. *Time, cells, and aging*. Academic Press, N.Y.

Sussman, R. R., and M. Sussman. 1963. Ploidal inheritance in the slime mould *Dictyostelium discoideum*: Haploidization and genetic segregation of diploid strains. *J. Gen. Microbiol., 30*:349-355.

Takeuchi, I. 1960. The correlation of cellular changes with succinic dehydrogenase and cytochrome oxidase activities in the development of the cellular slime molds. *Develop. Biol., 2*:343-366.

Takeuchi, I. 1963. Immunochemical and immunohistochemical studies on the development of the cellular slime mold *Dictyostelium mucoroides*. *Develop. Biol., 8*:1-26.

Tartar, V. 1961. *The Biology of Stentor*. Pergamon Press.

Tartar, V. 1962. Morphogenesis in Stentor. *Adv. in Morphogen., 2*: 1-26.

Tinbergen, N. 1951. *A Study of Instinct*. Oxford University Press.

Tinbergen, N. 1953. *Social Behavior in Animals*. Methuen, London.

Thompson, D'A. W. 1942. *Growth and Form*. Cambridge University Press.

Torrey, J. 1963. Cellular patterns in developing roots. In *Cell Differentiation*, ed. by G. E. Fogg (17th *Soc. Exp. Biol.* Symposium) Cambridge University Press.

Waddington, C. H. 1940. *Organizers and Genes*. Cambridge University Press.

Waddington, C. H. 1957. *The Strategy of the Genes*. George Allen and Unwin, London.

Weismann, A. 1891. *Essays upon Heredity and Kindred Biological Problems*. 2nd ed. Oxford University Press.

Wheeler, W. M. 1911. The ant colony as an organism. *J. Morph., 22*:307-325.

Wheeler, W. M. 1922. *Social Life among the Insects*. Constable, London.

White, M. J. D. 1945. *Animal Cytology and Evolution*. Cambridge University Press.

Wilson, C. M. 1952. Meiosis in *Allomyces*. *Bull. Torrey Bot. Club, 79*:139-160.

Wilson, E. B. 1924. *The cell in Development and Heredity*. 3rd ed. Macmillan, N.Y.

Wittingham, W. F., and K. B. Raper. 1960. Non-viability of stalk cells in *Dictyostelium*. *Proc. Nat. Acad. Sci., 46*:642-649.

Wynne-Edwards, V. C. 1962. *Animal Dispersion in Relation to Social Behavior*. Oliver and Boyd, Edinburgh.

Index

Acetabularia, 34, 101; fragmentation in, 54
Achlya, spores of, 64
Ackert, J. E., 95, 96
acrasin, 89
Acytostelium, 57; size decrease in, 54
adhesiveness, of cells, 14
Adolph, E. F., 29, 95
Agalma, Plate 28
age group mixing, in populations, 47, 48
aggregations, role in the regulation of animal numbers, 113, 135
aggregative organisms, 30-32, 89-90; size decrease in, 54
alarm notes, 134
Alexopoulos, C. J., 108
algae, *Plates 8, 10, 11, 12*; alternation of generations in, 159; filamentous, fragmentation in, 54, 56; spores of, 64
allometry, 192-194
Allomyces, Plate 13; dormancy in, 173, 174; progressive cleavage in, 56
alternation of generations, 159, 160; relation to selection, 132
ammonites, Cope's law and, 179
amoeba, *Plate 2*; differentiation in time, 101; fission in, 53; synthesis of DNA in, 27
Amoeba proteus, reduced weight in, 29
amphibians, metamorphosis and size, 21; neoteny in, 122; size, decrease in, 126; size, in evolution, 177; size *vs.* generation time, 16, 17
analogy *vs.* homology, 9
anastomosis, in fungi, 45
angiosperms, incompatibility mechanisms in, 153; reduction of gametophyte, 132
annelids, fission in, 53
antlers, size and allometry, 192
ants (*see also* army ants), 49, *Plate 30*; speed of, 186
aphids, dormancy in, 175; parthenogenesis in, 60, 145
apomixis, 145
Aristotle, 5

armadillos, quadruplets in, 55
army ants, 49; colony splitting in, 53, 106, 107
arthropods, Cope's law and, 179; size equilibrium in, 69
Ascaris, chromosome behavior in, 86
ascidians, *Plate 24*; bisexuality in, 153; colonial development of, 45; metamorphosis of, 21, 132; mosaic eggs of, 92, 93; regeneration in, 85; size decrease in, 61, 126
ascomycetes, dormancy in, 174; haploidy in, 158; recombination in, 175
ascospores, 64
asexual buds, in size decrease, 57
asexual organisms, 143ff.
asexual spores, in size decrease, 60
Aspergillus, 36; size of, 68
auxin, 83, 112
auxopores, in diatoms, 181
Avery, O. T., 150
Axolotl, neoteny in, 122

Bacillus, Plate 1
bacteria, *Plate 1*; adaptations in, 142; differentiation in time, 101; dry mass, increase in, 29; fission in, 53; generation time of, 29, 196, 197; niches of, 190; precursors of eucells, 129, 130; production of multinucleate filamentous forms, 33; size increase in, 27, 29; size *vs.* generation time, 16, 17; speed of, 185; spores of, 64; structure of, 27, 29
Balanoglossus, relation to echinoderms, 133
Baldwin effect, 79
Barth, L. G., 93
Bartnicki-Garcia, S., 33
basal granules, in ciliates, 91
basidiomycete, 38, *Plate 14*; anastomosis in, 45; haploidy in, 159; incompatibility mechanisms in, 153
basidiospores, 64
bats, delayed implantation in, 65; speed of, 187
Beale, G., 142
Beermann, W., 76
beneficial death, 170
Bernard, C., 81, 195

{ 211 }

Commoner, B., 76
communication, 134; between cells, 83; between cells in aggregative organisms, 32; between organisms, 47, 84; in bird flocks, 90; in fertilization, 117, 118; in orientation to mate, 154; in rigid cell masses, 94
community, stability of, 136
complexity, definition of, 18; correlation with absence of sexuality, 146; relation to diploidy, 157; relation to locomotion, 41; relation to stability, 136
conifer, size in evolution, 177
conjugation, in ciliates, 91
Conklin, E. G., 92, 93
contact adhesion, 89
convergence, in evolution, 9, 165
Cope, E. P., 178
Cope's law, 178-181, 192
Coprinus, Plate 14; size of, 68
coral, absence of senescence in, 170
cordaites, size in evolution, 177
Costello, D. P., 93, 133
cricetines, rate of evolution of, 182
cross breeding, favoring of, 152-155
cross-veinless, in *Drosophila*, 80
Crowell, S., 46, 62, 69, 110
crystals, analogy to identical genomes, 82, 88
cysts, in unicellular organisms, 29, 64
cytoplasmic inheritance, 77; cortex in ciliates, 92

Dalcq, A., 93
Daniel, J. W., 108
Darlington, C. D., 149, 153
Darwin, Charles, 5
day length, as a stimulus, 118
de Beer, G. R., 5, 119-125, 132
deermice, speed of, 186
delayed implantation, 65
developmental noise, 95, 99
diapause, 23, 42, 66; factors causing termination of, 114
diapho-epiphysial junction, 112
diatoms, size decrease in, 178-181
Dictyostelium (*see also* cellular slime molds), *Plate 6*; division of labor in, 57; germination in, 117
differentiation, definition of, 26; by changes in time, 100; due to environment, 93; in mobile units, 94ff.; in rigidly spaced units, 91ff.; spatial, in unicellular organisms, 101, 102

diffusion forces, in ascidians, 93; in spacing, 73
Diplodocus, size in evolution, 177
diploidy, 155ff.; relation to complexity, 157
diplophase, role in life cycle, 8, 43
Dippell, R. V., 109
displays, role in regulation of animal numbers, 113
division of labor, 165; in time, 100ff.
DNA, 76, 77; and synchronous divisions, 105; as an index of cell size, 27; in chromosomes, 150; synthesis in bacteria, 27
Dobzhansky, Th., 191
dolphin, speed of, 185
dormancy, advantages of, 171; cases of, 23
doves, gametogenesis in, 109
dragonfly, speed of, 187
Drosophila, speed of, 187
dry mass as a measure of cell size, 27, 28
Dugesia, Plate 20; fission in, 53
dwarfism, pituitary control of, 112

ecdysone, inducer of meiosis, 145
echinoderms, Cope's law and, 179; metamorphosis in, 21, 59, 126, *Plate 23*; relation to chordates, 133
ectohormones, 86, 98
egg, as point of minimum size, 21; as resistant stage, 23, 65
elephants, rate of evolution of, 182
Elton, C. S., 190, 191
Emerson, A. F., 170
Emerson, R. E., 173, 174
endocrines, in relation to behavior, 109; in size control, 112
energids, 101
Enteromorpha, 36, 37
enzymes, as reactants in steps, 72
Ephrussi, B., 77
Ephydatia, 116
epigenesis, definition of, 75
epigenetic canals, 122
Escherichia coli, generation time of, 197
Eucapsis, 36, 37; polarity in, 90
eucells, definition of, 27; generation time of, 29; minimum requirements of, 129; synthesis of DNA in, 27
Euglena, interval rhythms of, 79
eurypterids, size in evolution, 177
evolution, rate and population size, 191

mitotic spindle, as an index of growth direction, 35, 36
modifying genes, 134, 137
molluscs, bisexuality in, 153; size *vs.* generation time, 16, 17
Moore, J. A., 85
Morgan, T. H., 153
morphogenetic movement, definition of, 26; in aggregative organisms, 30-32; in myxomycetes, 34
Morton, A. G., 107
mosaic development, 40, 92
moss, 8, 43, *Plate 15*; alternation of generations in, 60; dormancy in, 174
motility (*see* locomotion)
Mucor, colonial development of, 44; haploidy in, 157; size of, 68, 69; spatial *vs.* temporal differentiation in, 102; spores of, 64; uni- and bisexuality in, 154
multicellular organisms, 34ff.
multinucleate organisms, 32-34; differentiation in time, 100, 101; fragmentation in, 54
multiple choice variation, 140ff.
muscle, contraction in relation to size, 188
mushrooms, 38, *Plate 14*; colonial development of, 44; size of, 68
mutation, in chains of steps, 123; in relation to the life cycle, 12
Myriophyllum, 140, 141
myxobacteria, 30, *Plate 5*; differentiation in time, 101; polarity in, 89; recombination in, 147; size decrease in, 54
Myxococcus, size decrease in, 54
myxomycetes, 33, 108, *Plate 7*; differentiation in time, 100; progressive cleavage in, 54; resting stage, 66; spore formation in, 107; spore germination in, 115, 117

natural selection, invariant structures and, 129; relation to steps, 11-13
Necturus, neoteny in, 122
neoteny, 120-122
Neurospora, colonial development of, 44; haploidy in, 158; spatial *vs.* temporal differentiation in, 102; spermatia of, 65
Newell, N. D., 178, 179
niacin, in myxomycete fruiting, 108
niches, in relation to size, 190
Nickerson, W. J., 33

Nipkow, H. F., 181
nonselectable structures, 129-131
nucleic acids, as ubiquitous chemicals in cells, 130
nucleus-cytoplasm ratio, 106
number control, in populations, 48

Obelia, growth and regression cycles in, 46, 62, 69, 110
oöplasmic segregation, 93, 94
opossums, rate of evolution of, 182
organ forming substances, 93
organism recombination, 151ff.
origin of life, 74
ostrich, speed of, 186
Oxytricha, cannibalism in, 184

paedogenesis, 120, 121
Pandorina, cell polarity in, 90
Papazian, H., 172
Paramecium, 33; inbreeding–outbreeding strains in, 154, 155; multiple choice variation in, 142; range variation in, 95-99; senescence in, 109
parasites, niches of, 190; size equilibrium in, 66
Pardee, A. B., 142
parthenogenesis, 115, 145; in mosses, 160
Pavan, C., 76
peck-order, 98, 99, 113
Pediastrum, size of, 67, 68
penguin, swimming speed of, 185
Penicillium, 36; colonial development of, 44; parasexuality in, 148; spore formation in, 107
phage, generation time of, 196
phenotypic variation, 140ff.
phycomycetes, dormancy in, 173, 174; progressive cleavage in, 56; spores of, 64
phylogeny, in classification, 6
Physarum, *Plate 7*
physiological cycles, 23, 24; in relation to the life cycle, 12, 13
Picken, L. E. R., 14, 27, 129
Pittendrigh, C. S., 79
pituitary, growth hormone, 112
planarians, 106, *Plate 20*; fission of, 14; regeneration in, 85
plants, unisexuality in, 153
plasmogamy, 158
pleiotropy, 77, 126, 137, 163
pluteus larva, metamorphosis of, 42